Christopher Columbus, Paul Leicester Ford

Writings of Christopher Columbus

Descriptive of the discovery and occupation of the new world

Christopher Columbus, Paul Leicester Ford

Writings of Christopher Columbus
Descriptive of the discovery and occupation of the new world

ISBN/EAN: 9783337164218

Printed in Europe, USA, Canada, Australia, Japan

Cover: Foto ©berggeist007 / pixelio.de

More available books at **www.hansebooks.com**

WRITINGS OF
CHRISTOPHER COLUMBUS

DESCRIPTIVE OF

THE DISCOVERY AND OCCUPATION
OF THE NEW WORLD

EDITED, WITH AN INTRODUCTION,

BY

PAUL LEICESTER FORD

New York
CHARLES L. WEBSTER & CO.
1892

TO
E. W. BLATCHFORD
OF CHICAGO

AS A TRIBUTE OF RESPECT BOTH TO HIMSELF
AND TO HIS CITY

SO APPROPRIATELY AND NOBLY
COMMEMORATING

CHRISTOPHER COLUMBUS

THIS SMALLER MEMORIAL
IS DEDICATED

CONTENTS

	PAGE
INTRODUCTION	11
LETTER TO FERDINAND AND ISABELLA	26
LETTER TO RAPHAEL SANCHEZ	33
LETTER TO LUIS DE SANTANGEL	52
LETTER TO FERDINAND AND ISABELLA	67
PRIVILEGES OF COLUMBUS	75
DEED OF ENTAIL	83
LETTER TO FERDINAND AND ISABELLA	105
LETTER TO JUANA DE LA TORRES	151
PRIVILEGES OF COLUMBUS	177
LETTER TO FERDINAND AND ISABELLA	199
WILL OF COLUMBUS	240

INTRODUCTION

In the four centuries which have elapsed since Christopher Columbus sailed his caravels to the westward in search of the Indies, and found in their stead a new world, two factions have developed, holding antagonistic opinions regarding his personality and achievements. The one, regardless of flaws in his character and methods, seeks his canonization from the great Church of which he was so ardent a follower; claiming that the conception, accomplishment, and results of his great discovery entitle him to a place among those whom reverent mankind must worship, as guided by more than earthly and human inspiration. The other, losing all breadth of view in the minuteness of its investigation of his life and works, finds him a vain, ignorant, and even half-mad enthusiast, his claim of Biblical prophecy and inspiration from the Trinity little better than blasphemy, his great act based on ignorance and error, and the result nothing but a lucky chance. Nor need these views occasion surprise. Such opposite opinions of the same person are not uncommon either of those who have made or are making history. As long as there are two ends to our mental opera-glass we shall have varying ideas of the magnitude of men and deeds, depending on whether the magnifying or diminishing view is taken. The matter for surprise is that each faction—the "big Endians and the little Endians"

—should not realize that there are two ends; that a man, as well as a statue, can be regarded from above as well as below, and that though the two impressions are very different, yet the man is the same; that the two, or many views must be combined to produce a true whole, and that either by itself is misleading and untruthful. Thus in the life of Columbus there is much to sustain both factions, yet the view of each conveys a false result. To critically discuss the basis for these two extremes is neither possible nor desirable here. But a few words as to how far we are indebted to Columbus for the discovery of the western world will serve to indicate the great truths in both points of view, as well as their wrong impression.

The theory of the rotundity of the earth naturally carried with it the corollary that, by sailing westward across the Atlantic, land would eventually be reached in the east, and this opinion was therefore held by Aristotle, Pliny, Marinus, Strabo, and other ancient writers. But this knowledge, except as a theory, was of no value. To Greece or Rome the eastern trade was too small and too easily conducted to tempt explorations of new routes for it. Nor did suggestions of unknown lands beyond the western limits of Europe arouse curiosity or desire to explore them. In vain did Theopompus, four centuries before Christ, and after him Virgil, Plato, Aristotle, Seneca, and many others, write of such. Europe itself was too new and had too little energy to spare, to undertake the mere verification of what at best were mere theories. Before Seneca's marvelous prophecy that "there will come an age in which Ocean shall loosen the bands of things; a great country shall be discovered . . . and Thule shall no longer be the extremity of the earth," could be more than a prediction, population or trade must press for new outlets. Barring accidents, on these causes depended the nature of the exploration of the Atlantic. Should population first crowd, the mythical lands would be first sought. Should trade need new routes, then Europe would seek them in the vast unknown waters beyond the Pillars of Hercules. Till in want of one or the other, the opinions and hypotheses of these writers would be but interesting speculations for the learned; not matters for practical men to waste time on. Yet, if the early hypotheses produced no direct results, they nevertheless became later of much importance, for the respect given to the classic writers induced many to accept their opinions, who would not have been convinced by the more practical evidence which later times produced.

For over a thousand years Europe was busied with the work of settling, holding, and consolidating its territories into countries strong enough to insure both self-preservation and progress. And this process involved a destruction of population that prevented all necessity for new lands. Had the exploration of the Atlantic depended on this alone it is probable that the western continents would have remained unexplored at least two centuries longer. While Europe, however, felt no pressure of population and therefore gave no heed to the tales and prophecies of unknown lands, the other theory, of reaching the east by sailing westward, became, by circumstances, of more importance. From the earliest time a trade had been carried on between

Europe and Asia. During the semi-barbaric and feudal times which succeeded the Roman empire this was of especial value to the former, for the peace and knowledge needed for arts and manufactures were slow to come, and it turned to the older civilization for them, sending in exchange such few things as it possessed that Asia wanted, and eking out its balances with gold and silver. So steadily had these been drained away from Europe during these stormy centuries that Asia was termed by it "the Golden East." The trade with this land of jewels, precious metals, wondrous fabrics, and spices, was the most profitable of the time, and enriched the centres of distribution to a marvelous degree. First Constantinople, then Amalfi, and finally Venice and Genoa, became rich and great by it. The routes through the Black Sea, Persian Gulf, and Red Sea, though involving caravans and transshipment, were sufficient; and from the Mediterranean the products were distributed over Europe with ease. Europe stood with its back to the west. All the interest and thought it could spare from its own concerns were centered in the east, and but for an accident would have so continued for an indefinite time. Yet so curiously are causes and effects blended that it was this very concentration and intensity of interest that eventually led to the exploration of the great west.

The cause for this apparent contradiction is clear and obvious. In the ninth century, a new element entered into the problem of barter between the civilized east and the fast civilizing west. Between these two, and walling them from each other, the barbaric Turk threw himself. Across the caravan routes of Asia

Minor, Arabia, Palestine, and Egypt he pushed his hordes. The trade was too valuable to Europe to be yielded without a struggle. For four hundred years, with a curious, perhaps unconscious, combination of religion and pelf, it struggled by the crusades to crush this menace. In vain. Slowly but surely, the trade was throttled. And these very crusades stimulated the demand for eastern goods, for it familiarized the more extreme and northern nations of Europe with them. So strong did this demand become, that, force having been tried in vain, even religion was sacrificed, and the Pope gave dispensations which allowed Christian nations to make treaties and leagues with the hated Mohammedans. Even this availed not to preserve the trade. On land the Turk could not maintain order enough to make caravan trading other than a most hazardous venture, and Turkish pirates scoured the eastern end of the Mediterranean, capturing without heed of treaty. If Europe wished the products of the east, it must find new routes for them.

This turned European thought in new directions, and set it considering and discussing the Atlantic. The old writers were studied and quoted. New ones added their opinions, and the few facts that had been learned in the intervening time. The growth of interest in the east had led to occasional travelers visiting that region, and thus a very fair knowledge, for the times, of the extent of it had been obtained. Something, too, had been learned of the eastern coast of upper Africa. The problem was to learn whether it was shorter to circumnavigate Africa, and so reach the Indies, or to sail directly westward. This practically depended on the size of the earth and on the extent of southern Africa, and as to both, ancient and modern philosophers differed. Except as an opinion, no conclusion could be reached without actual experiment. The nations most likely to attempt this were the Italian republics, which so greatly profited by this trade; but they clung to the old and accustomed routes to the last, not being able to read the handwriting on the wall. Nor did the northern nations, hardy and venturesome as their sailors were, make the endeavor. Furthest from the east, the benefits of the trade to them were too slight to be appreciated. But midway between these, the nations of the Spanish peninsula, facing both east and west, and seeing the richness of the oriental trade, yet only seeing it, were, *a priori*, the countries to seek for new routes to " far Cathay." And so it proved.

With the small ships, and the imperfect means of navigation, the African route was almost certain to be the first attempted, as allowing the ships to hug the land. And to this Portugal turned her energies. Expedition after expedition crept down that coast, every few years going a little further than before; till it became evident that that route was far longer than had been hoped, and the question of a possibly shorter one directly across the Atlantic became more important. Disappointed, yet not despairing, Portugal clung to the former, and so to Spain fell the honor of exploring the latter.

The man to act for Spain was singularly fitted for the work. Columbus, though calling himself " an ignorant man," had just the needed knowledge. Born in Genoa, his attention could not be otherwise than attracted to the Indies, and to the valuable trade with them which made the little republic a power. Whether he ever was at the University of Pavia has been questioned, on grounds that seem altogether insufficient, as they consist only of his extreme youth, entirely ignoring the precocity of the southern nations, and the early age at which lads were then thrown out in the world. Certain it is that later we find him with a good knowledge of Latin and a reader of books, so that such aid as was to be found in the learning of his time was not barred to him as it was to many of that day. At fourteen he turned mariner, as was natural to a Genoese, and, to use his own words, "followed the seas for twenty-three years without being on shore any space of time worth accounting." Before 1492, he had been to the Grecian Archipelago, Guinea, the Canaries, England, and Iceland, and indeed he claimed to have seen all the East and West, and that "wherever ship had sailed, there have I journeyed." Without entirely abandoning the sea he then became a map-maker, and to such it seems scarcely possible that the theory of a westward sailing should not have occurred, even if no ancient writer had taught it, and if it had not been a matter of constant and growing discussion. Thus a man of bold disposition, good mind, and education, with practical sea and cartographic knowledge, and withal an enthusiast, was ready to act at the time that Europe's interest in the east forced her at last to turn westward. That he had to beg and plead with different countries proves that he, like most men of single ideas, was in advance of the world's progress. That he offered his services to Genoa and Venice, though denied by many modern writers, on purely negative evidence, is so natural as to be well-nigh certain. The evident cause for their refusals has already been indicated. So, too, has Portugal's reason for declining his offer. And so, the "Most Christian, High, Excellent, and Powerful Princes, King and Queen of Spain and of the Islands of the Sea . . . in consequence of the information which I had given your Highnesses respecting the countries of India . . . determined to send me, Christopher Columbus, to the above mentioned countries, . . . and furthermore directed that I shall not proceed by land to the east, as is customary, but by a westerly route." And sailing from Palos on Aug. 3, 1492, he found, not a trade route to the Indies, but a new world.

The question whether Columbus first discovered this world has been an endless and decidedly amusing bone of contention among historians and pseudo-historians. Every myth or shadowy suggestion of a possible pre-Columbian finding of America has been brought forward, regardless of the basis for the claim. From almost every nation of Europe and Asia a pretender for the discovery has been exploited by some would-be historian and these have been soberly discussed even by more judicious writers, with much inevitable befogging of the whole question. Not till the unproved and non-disprovable stories of the Egyptian, Arabian, Phœnician, Tartar, Chinese, Irish, Welsh, Venetian, Portuguese, and Polish discoveries of the new world are dismissed into the lumber-room of history, can the true value of pre-Columbian progress toward the finding of the new world be realized by any but specialists.

Rejecting them, certain facts stand forth with clearness.

The first of these is the Norse finding and attempted occupation of America. Few historical facts deserve more credence. Unimpeachable records show that in the seventh century political refugees from Norway occupied Iceland. In the next century, by chance, Greenland was found and colonized; and in the ninth century, again by chance, the American continent was stumbled upon. Settlements were attempted, but in vain. Europe had too little need for new countries, and too much need for men, to spare enough for them to hold their own against the natives. For three hundred years occasional voyages were made to the coast, and the settlement at Greenland outlasted even these; not being crushed till the century in which the new world was found by the Spaniards.

No such records exist concerning any other accidental finding of America, yet from certain data of equal credibility, the proof that various parts of the western islands and continents were occasionally sighted by stray ships, is so strong that the evidence should be sufficient to prove the claim in a court of justice. Historians, for the most part, and the biographers of Columbus, have denied this, basing their denials on the absence of records of such. It is true that, so far as we know, no books or chronicles of the fourteenth or fifteenth centuries contain mention of such voyages. But this is purely negative testimony, and is not the evidence to cite. With hardly an exception, the writers of that time were secluded in the walls of monasteries, out of touch with the busy world, and shut off from and even despising current and mundane news. As soon expect to find references to voyages and strange lands in their writings, as the latest stock quotations in the modern university or cathedral. On the wharves of commercial cities, and in the caracks and caravels that lay beside them, was where the news of unknown lands was to be heard. Here the air was full of stories and rumors, which have left so strong an atmosphere of knowledge of western lands, that it has given pause even to the most ardent believers in the priority of the discovery of Columbus, and has driven them to confess that stories of western lands were "in the air" long before Columbus sailed, but they claim that the only basis for these rumors and stories was mariners' yarns, imaginings, and deceptions. Were this atmosphere the only proof, their assertions might be granted, as these sailors' tales found no chronicler to verify and preserve them. But they come down to us in another form, which cannot be dismissed so easily, and which as yet seems by far too little studied and appreciated—the evidence of the cartography of those centuries.

The map-maker of that time stands on a plane with, yet apart from, his fellow-scholars, being the only branch which was called upon to keep down to date. A man of learning, he was, as well, one who had usually commanded or sailed in ships, and he lived in the seaport towns where he could find a sale for his charts, and obtain the latest facts of geographical discovery. The more lands he could map accurately, the greater his repute, and the larger his profits. That this would make him careful and painstaking was certain. Constructing maps by which vessels were to be navigated, he would not idly enter lands to obstruct routes, nor would he omit from his chart any reported on good evidence lest he should cast ships away on them. Of his superior information on this subject to his contemporaries, it is enough to cite that he accurately mapped the Canaries, Madeira, and the Azores, respectively ninety-six, fifty-one, and ninety-eight years before those islands were discovered by the Portuguese. The most accurate and careful maps of to-day are the Admiralty and Coast Survey charts, and this was equally true of the maps of five hundred years ago. Their evidence cannot be omitted in the history of the discovery of America.

The persistence with which lands unknown to the learned of Europe were entered on the maps of the fourteenth and fifteenth centuries is really remarkable. In the few maps that have been preserved to us, no less than twenty-seven, made between 1351 and 1492, by different geographers working in different cities of Europe, locate islands or masses of land in the western Atlantic, of which no account or mention is to be found in the writings of the same time. Varying in size and spelling, Greenland, Brazil, Antilia, St. Brandan, Roillo, Satanaxio, Sete Zitade, Saliroza, and others, are given. Sometimes only one is included; sometimes several. That their relative positions were varied on different maps and that different names were given to the same lands, has been cited as evidence that they were the mere coinings of the different draftsmen's imagination. But the purpose of these maps, as already shown, proves the unlikelihood of this. And this very contradiction really increases the proof of their reality, for had they been only inventions, it is obvious that their positions would have remained unchanged from the places assigned them by their inventor, while it is evident that each mariner who reported a land-fall to westward would give it a name and report the latitude. And it is not strange that, unacquainted with the fact that two continents with thousands of miles of coast existed, the map-makers should move these islands up and down in the Atlantic, and confuse names in the attempt to harmonize and reduce these reports to cartography. Of the sailors' faith in the truth of these maps, it is enough to mention that the chart by which Columbus sailed in his first voyage included some of these islands; and that he believed them well-established facts is shown by his narrative of this voyage, in which he several times speaks of them. Even after the finding by Columbus, and the then rapid exploration of the coast, it was many years before America was not usually mapped as a series of islands, often varying in name, each of which represented a stretch of territory described by some explorer or explorers, and which the ignorance of the times prevented uniting into one continuous coast-line.

That these pre-Columbian land-falls produced no stir or results has been urged as a reason for their never having been made. The cause is clear. Europe did not want new lands and cared nothing for them. Many of the sailors' stories, though believed by their fellow-mariners and map-makers, received no credence from courts or colleges. Even a monarch so interested and learned in geographical discovery as John of

Portugal, refused to believe the finding of new lands by Columbus, till he was shown the natives and found them a race hitherto unknown. The failure of the Norsemen, a nation of seamen and possessing the shortest route to America, to take advantage of their knowledge, proves how valueless these discoveries were then considered. For nearly two hundred years after America was known, only deportations, greed of gold, or political or religious persecutions could force the peoples of the old world to migrate to the new. That the voyage of Columbus produced more results than these other accidental findings was due to three causes: 1. That it was made under the patronage of a court which anthenticated, advertised, and shared the glory of the discovery. 2. That for many years the lands found were believed to be the East Indies, in which Europe was greatly interested. 3. That Columbus on his return reported vast mines of gold, thereby inciting many to journey westward.

That the discovery of America by Columbus was as accidental as those of the Norsemen or the unrecorded findings, and that he died ignorant of the fact that he had found a new world, are urged by many as militating against, if not depriving him of, the claim to its honor. Such arguments entirely overlook the fact that Columbus was the first purposely to attempt the exploration of the western Atlantic, and that whatever results this exploration produced are due to him. Nor are the great results the only honor that must be accorded him. A man who, for twenty years brooded on a great and useful idea, who battled with all forms of human opposition to that idea, and who risked his life in unknown lands and waters in leaky and unfit ships, with untrustworthy subordinates, to prove that idea— even if we find him boasting vainly of his great deeds, or whining in humiliation and chains—must challenge the admiration of the world, as an advanced thinker and a brave man. The defects delighted in by his "critical" biographers and commentators undoubtedly existed, he being human. But self-interest and self-esteem are not such rare qualities as to form articles of impeachment against the men who have correspondingly great merits. Other faults were those of his generation. To animadvert on his claim of divine inspiration for his westward sailing; on his belief that he had discovered the locality of the Garden of Eden; on his theory of a pear-shaped world; on his eager searches for gold; on his proposed enslavement of the aborigines, and on his bastard child and its mother, is anything but "critical," for it is projecting the atmosphere and views of the skeptical, Protestant, nineteenth century, into the believing, Catholic, fifteenth century.

The best antidote to these views, the editor believes, is to be found in the writings of Columbus. Only fragments, with wide gaps and breaks, due to loss and destruction, they nevertheless show, as nothing else can show, the thoughts, acts, and desires of the man. He himself once entreated, little recking that it would be equally necessary nearly four hundred years later, that "I must be re-established in reputation, and spoken of throughout the universe: for the things I have done are such, that they must gain, day by day, in the estimation of mankind." If this little volume contributes at all to that wish, the purpose of its editor has been accomplished. PAUL LEICESTER FORD.

Writings of Columbus

LETTER TO FERDINAND AND ISABELLA*

IN the name of our Lord Jesus Christ. Whereas, Most Christian, High, Excellent and Powerful Princes, King and Queen of Spain and of the Islands of the Sea, our Sovereigns, this present year, 1492, after your Highnesses had terminated the war with the Moors

---

*This narrative of the first voyage of Columbus is usually quoted by historians as his "Journal," and they deplore the loss of the letter he wrote to the King and Queen of Spain recounting his discoveries. But a very superficial study of it must convince anyone, from the use of the personal pronoun, that this is that letter, and not a journal, though Columbus adopted the diary form in writing it. Unfortunately, only an abstract of the original, made by Las Casas, is known to us, and he so changed and abbreviated the words of Columbus, that the bulk of it is in no sense the latter's writing. The preamble, however, Las Casas copied entire, so it is here printed. This portion of the letter was probably written the day before his sailing from Palos, Aug. 3, 1492. The translation is by Samuel Kettell, printed in his *Personal Narrative of the First Voyage of Columbus*, . . . *Boston*, 1827. The original text is in Navarrete's *Coleccion de los Viages*, . . . *Madrid*, 1825.

reigning in Europe, the same having been brought to an end in the great city of Granada, where, on the second day of January, this present year, I saw the royal banners upon the towers of the Alhambra, which is the fortress of that city, and saw the Moorish king come out at the gate of the city and kiss the hand of your Highnesses, and of the Prince, my Sovereign; and in the present month, in consequence of the information which I had given your Highnesses respecting the countries of India and of a Prince called Great Can,* which in our language signifies King of Kings, how, at many times, he and his predecessors had sent to Rome soliciting instructors who might teach him our holy faith, and the holy Father had never granted his request, whereby great numbers of people were lost, believing in idolatry and doctrines of perdition. Your Highnesses, as Catholic

* This refers to Kooblai Khan, the great founder of the Mongol empire. Marco Polo had exploited his personality, wealth, and empire, to Europe, a century before, and Europe persistently kept him alive for long after even this period.

Christians, and princes who love and promote the holy Christian faith, and are enemies of the doctrine of Mahomet, and of all idolatry and heresy, determined to send me, Christopher Columbus, to the above-mentioned countries of India, to see the said princes, people and territories, and to learn their disposition and the proper method of converting them to our holy faith; and, furthermore, directed that I should not proceed by land to the East, as is customary, but by a Westerly route, in which direction we have hitherto no certain evidence that any one has gone. So, after having expelled the Jews from your dominions, your Highnesses, in the same month of January, ordered me to proceed, with a sufficient armament, to the said regions of India, and for that purpose granted me great favors, and ennobled me that thenceforth I might call myself Don, and be High Admiral of the Sea, and perpetual Viceroy and Governor in all the islands and continents which I might discover and acquire, or which may hereafter be discovered and acquired in the ocean; and that this dignity should be inherited by my eldest son, and thus descend from degree to degree forever. Hereupon I left the city of Granada, on Saturday, the twelfth day of May, 1492, and proceeded to Palos, a seaport, where I armed three vessels, very fit for such an enterprise* and having provided myself with abundance of stores and seamen,† I set sail from the port, on Friday, the third of August, half an hour before sunrise, and steered for the Canary Islands of your Highnesses, which are in the said ocean, thence to take my departure and proceed till I arrived at the Indies, and perform the embassy of your Highnesses to the Princes‡ there, and

---

* These were the *Santa Maria*, *Pinta* and *Nina*. Columbus's opinion of their fitness very quickly changed after he got to sea. Unlike the Portuguese caravels, the Spanish caravels were square rigged, and with the high bows and poops of those days they were practically incapable of beating to windward. His largest ship, too, was a very dull sailor, and all three leaked so as hardly to be safe.

† There is much contradictory evidence as to how many men sailed with Columbus, but probably Peter Martyr's statement of two hundred and ten is the most reliable.

‡ The monarchs had given him a letter in blank to them,

discharge the orders given me. For this purpose I determined to keep an account of the voyage, and to write down punctually every thing we performed or saw from day to day, as will hereafter appear. Moreover, Sovereign Princes, besides describing every night the occurrences of the day, and every day those of the preceding night, I intend to draw up a nautical chart, which shall contain the several parts of the ocean and land in their proper situations; and also to compose a book to represent the whole by picture with latitudes and longitudes, on all which accounts it behoves me to abstain from my sleep, and make many trials in navigation, which things will demand much labor.*

which is printed in Vol. I. of the *Calendar of State Papers Relating to England and Spain.*
  *On his return from his first voyage Columbus gave this to the Queen, who wrote him in September, 1493: " By this courier I send you a copy of the book which you left here. The reason of its being so long delayed, was to have it written out secretly, in order that neither the Portuguese who are here, nor any other person, might know anything of it. And for that purpose, that it might be more quickly finished, you will perceive that it is in

two different handwritings. Certainly, according to what has been here seen and treated of respecting this affair, we perceive every day, more and more, its great weight and importance. If the sailing chart which you were to prepare is finished, send it to me immediately."

LETTER TO RAPHAEL SANCHEZ*

Letter addressed to the noble Lord Raphael Sanchez, Treasurer to their most invincible Majesties, Ferdinand and Isabella, King and Queen of Spain, by Christopher Columbus, to whom our age is greatly indebted, treating of the islands of India recently discovered beyond the Ganges, to explore which he had been sent eight months before under the auspices and at the expense of their said Majesties, which the noble and learned man, Aliander de Cosco, translated from the Spanish idiom into Latin the third day of the calends of May, 1493. The year One of the Pontificate of Alexander VI.

KNOWING that it will afford you pleasure to learn that I have brought my undertaking to a successful termination, I have decided upon writing you this letter to acquaint you with all the events which have occurred in

* This letter, which is that which announced to Europe the discovery of Columbus, is unknown in the Spanish original. A copy of it was soon in Rome, and was there translated into Latin by Leander de Cosco, and in that language was many times reprinted in different parts of Europe. From this fact it is probable that it was received

my voyage, and the discoveries which have resulted from it. Thirty-three days* after my departure from Cadiz I reached the Indian sea,† where I discovered many islands, thickly peopled, of which I took possession without resistance, in the name of our most illustrious Monarch, by public proclamation and with unfurled banners. To the first of these islands,

---

by its recipient before the one Columbus addressed to Santangel, which I therefore place after this, though it is the far preferable account, as being translated directly from Columbus's words, instead of being a translation of a translation. The English translation is by R. H. Major, printed in his *Select Letters of Christopher Columbus*, . . . *London*, 1847. The original text is *Epistola Christofori Colom.* . . [*Rome*] *Maii. M. cccc. xciii.*

*This is an evident error, probably of the Latin translator. Major concludes that it was the careless substitution of Gadibus (Latin for Cadiz) for Gomera, in the Canaries, which was the last land Columbus visited before he sailed westward, and from which he was thirty-five days in sighting land. But this required two errors, and it seems to the present editor more probable that Columbus was in Cadiz two days before leaving Palos, and that the error was merely a typographical one in printing the Roman numerals XXXIII in place of LXXIII—an error far more likely to occur than the two changes of words involved in Major's explanation.

† October 12th.

which is called by the Indians Guanahani, I gave the name of the blessed Saviour (San Salvador), relying upon whose protection I had reached this as well as the other islands; to each of these I also gave a name, ordering that one should be called Santa Maria de la Concepcion,* another Fernandina,† the third Isabella,‡ the fourth Juana,§ and so with all the rest respectively. As soon as we arrived at that which I have said was named Juana, I proceeded along its coast a short distance westward, and found it to be so large and apparently without termination, that I could not suppose it to be an island, but the continental province of Cathay. Seeing, however, no towns or populous places on the sea-coast, but only a few detached houses and cottages, with whose inhabitants I was unable to communicate, because they fled as soon as they saw us, I went further on, thinking that in my progress I should certainly find some city or village. At length, after proceeding a great way, and

* North Caico.   ‡ Great Inagua.
† Little Inagua.   § Cuba.

finding that nothing new presented itself, and that the line of coast was leading us northward (which I wished to avoid, because it was winter, and it was my intention to move southward; and because, moreover, the winds were contrary), I resolved not to attempt any further progress, but rather to turn back and retrace my course to a certain bay that I had observed, and from which I afterwards dispatched two of our men to ascertain whether there were a king or any cities in that province. These men reconnoitred the country for three days, and found a most numerous population, and great numbers of houses, though small, and built without any regard to order: with which information they returned to us. In the meantime I had learned from some Indians whom I had seized, that that country was certainly an island: and, therefore, I sailed toward the east, coasting to the distance of three hundred and twenty-two miles, which brought us to the extremity of it; from this point I saw lying eastward another island, fifty-four miles distant from Juana, to which I gave the name of Española:* I went thither, and steered my course eastward, as I had done at Juana, even to the distance of five hundred and sixty-four miles along the north coast. This said island of Juana is exceedingly fertile, as, indeed, are all the others; it is surrounded with many bays, spacious, very secure and surpassing any that I have ever seen; numerous large and healthful rivers intersect it, and it also contains many very lofty mountains. All these islands are very beautiful, and distinguished by a diversity of scenery; they are filled with a great variety of trees of immense height, and which I believe to retain their foliage in all seasons; for when I saw them they were as verdant and luxuriant as they usually are in Spain in the month of May—some of them were blossoming, some bearing fruit, and all flourishing in the greatest perfection, according to their respective stages of growth, and the nature and quality of each: yet the islands are not so thickly wooded as to be impassable. The nightingale and various birds were singing in countless numbers, and

* Hispaniola, otherwise San Domingo or Hayti.

that in November, the month in which I arrived there. There are, besides, in the same island of Juana, seven or eight kinds of palm-trees, which, like all the other trees, herbs and fruits, considerably surpass ours in height and beauty. The pines, also, are very handsome, and there are very extensive fields and meadows, a variety of birds, different kinds of honey, and many sorts of metals, but no iron. In that island, also, which I have before said we named Española, there are mountains of very great size and beauty, vast plains, groves, and very fruitful fields, admirably adapted for tillage, pasture and habitation. The convenience and excellence of the harbors in this island, and the abundance of the rivers, so indispensable to the health of man, surpass anything that would be believed by one who had not seen it. The trees, herbage and fruits of Española are very different from those of Juana, and, moreover, it abounds in various kinds of spices, gold and other metals. The inhabitants of both sexes in this island, and in all the others which I have seen, or of which I have received information, go always naked as they were born, with the exception of some of the women, who use the covering of a leaf, or small bough, or an apron of cotton, which they prepare for that purpose. None of them, as I have already said, are possessed of any iron, neither have they weapons, being unacquainted with, and, indeed, incompetent to use them, not from any deformity of body (for they are well formed) but because they are timid and full of fear. They carry, however, in lieu of arms, canes dried in the sun, on the ends of which they fix heads of dried wood sharpened to a point, and even these they dare not use habitually; for it has often occurred, when I have sent two or three of my men to any of the villages to speak with the natives, that they have come out in a disorderly troop, and have fled in such haste at the approach of our men, that the fathers forsook their children and the children their fathers. This timidity did not arise from any loss or injury that they had received from us; for, on the contrary, I gave to all I approached whatever articles I had about me, such as cloth and many other things, taking nothing of theirs in return; but they are naturally timid and fearful. As soon, however, as they see that they are safe, and have laid aside all fear, they are very simple and honest, and exceedingly liberal with all they have; none of them refusing anything he may possess when he is asked for it, but, on the contrary, inviting us to ask them. They exhibit great love toward all others in preference to themselves; they also give objects of great value for trifles, and content themselves with very little or nothing in return. I, however, forbade that these trifles and articles of no value (such as pieces of dishes, plates and glass, keys and leather straps), should be given to them, although, if they could obtain them, they imagined themselves to be possessed of the most beautiful trinkets in the world. It even happened that a sailor received for a leather strap as much gold as was worth three golden nobles, and for things of more trifling value offered by our men, especially newly coined blancas, or any gold coins, the Indians would give whatever the seller required; as, for instance, an ounce and a half or two ounces of gold, or thirty or forty pounds of cotton, with which commodity they were already acquainted. Thus they bartered, like idiots, cotton and gold for fragments of bows, glasses, bottles and jars; which I forbade as being unjust, and myself gave them many beautiful and acceptable articles which I had brought with me, taking nothing for them in return; I did this in order that I might the more easily conciliate them, that they might be led to become Christians, and be inclined to entertain a regard for the King and Queen, our Princes and all Spaniards, and that I might induce them to take an interest in seeking out and collecting and delivering to us such things as they possessed in abundance, but which we greatly needed. They practice no kind of idolatry, but have a firm belief that all strength and power, and, indeed, all good things, are in heaven, and that I had descended from thence with these ships and sailors, and under this impression was I received after they had thrown aside their fears. Nor are they slow or stupid, but of very clear understanding; and those men who have crossed to the neighboring islands give an admirable description of everything they observed; but they never saw any people clothed, nor any ships like ours. On my arrival at that sea I had taken some Indians by force from the first island that I came to, in order that they might learn our language, and communicate to us what they knew respecting the country; which plan succeeded excellently, and was a great advantage to us, for in a short time, either by gestures and signs, or by words, we were enabled to understand each other. These men are still traveling with me, and although they have been with us now a long time, they continue to entertain the idea that I have descended from heaven; and on our arrival at any new place they publish this, crying out immediately with a loud voice to the other Indians: "Come; come and look upon beings of a celestial race;" upon which both women and men, children and adults, young men and old, when they got rid of the fear they at first entertained, would come out in throngs, crowding the roads, to see us, some bringing food, others drink, with astonishing affection and kindness. Each of these islands has a great number of canoes, built of wood, narrow and not unlike our double-banked boats in length and shape, but swifter in their motion; they steer them only by the oar. These canoes are of various sizes, but the greater number are constructed with eighteen banks of oars, and with these they cross to the other islands, which are of countless number, to carry on traffic with the people. I saw some of these canoes that held as many as seventy-eight rowers. In all these islands there is no difference of physiognomy, of manners, or of language, but they all clearly understand each other—a circumstance very propitious for the realization of what I conceive to be the principal wish of our most serene King, namely, the conversion of these people to the holy faith of Christ, to which, indeed, as far as I can judge, they are very favorable and well disposed. I said before that I went three hundred and twenty-two miles in a direct line from west to east, along the coast of the island of Juana; judging by which voyage, and the length of the passage, I can assert that it is larger than England and Scotland united; for, independent of the said three hundred and twenty-two miles, there are in the western part of the island two provinces which I did not visit; one of these is called by the Indians Anam, and its inhabitants are born with tails. These provinces extend to a hundred and fifty-three miles in length, as I have learned from the Indians whom I have brought with me, and who are well acquainted with the country. But the extent of Española is greater than all Spain from Catalonia to Fontarabia, which is easily proved, because one of its four sides, which I myself coasted in a direct line from west to east, measures five hundred and forty miles. This island is to be regarded with especial interest, and not to be slighted; for although, as I have said, I took possession of all of these islands in the name of our invincible King, and the government of them is unreservedly committed to his said Majesty, yet there was one large town in Española of which especially I took possession, situated in a remarkably favorable spot, and in every way convenient for the purposes of gain and commerce. To this town I gave the name of Navidad del Senor, and ordered a fortress to be built there, which must by this time be completed, in which I left as many men as I thought necessary, with all sorts of arms, and enough provisions for more than a year.* I also left them one caravel,† and skilful workmen, both in ship-building and other arts, and engaged the favor and friendship of the King of the islands in their behalf, to a degree that would not be believed, for these people are so amiable and friendly

---

* This, the first Spanish attempt to colonize the New World, resulted only in failure. On the return of Columbus, in his second voyage, he found that quarrels had arisen with the natives, and the Spaniards had been entirely exterminated.

† This is not clear. The *Santa Maria* was wrecked among the islands, was taken to pieces, and used to build the fortress at Navidad, and in that sense, and no other, was left there. The *Pinta* and *Nina* returned to Spain in company.

that even the King took a pride in calling me his brother. But supposing their feelings should become changed, and they should wish to injure those who have remained in the fortress, they could not do so, for they have no arms, they go naked, and are, moreover, too cowardly; so that those who hold the said fortress can easily keep the whole island in check without any pressing danger to themselves, provided they do not transgress the directions and regulations which I have given them. As far as I have learned, every man throughout these islands is united to but one wife, with the exception of the kings and princes, who are allowed to have twenty: the women seem to work more than the men. I could not clearly understand whether the people possess any private property, for I observed that one man had the charge of distributing various things to the rest, but especially meat and provisions, and the like. I did not find, as some of us had expected, any cannibals amongst them, but, on the contrary, men of great deference and kindness. Neither are they black, like the Ethiopians: their hair is smooth and straight, for they do not dwell where the rays of the sun strike most vividly—and the sun has intense power there, the distance from the equinoctial line being, it appears, but six-and-twenty degrees. On the tops of the mountains the cold is very great, but the effect of this upon the Indians is lessened by their being accustomed to the climate, and by their frequently indulging in the use of very hot meats and drinks. Thus, as I have already said, I saw no cannibals, nor did I hear of any, except in a certain island called Charis,* which is the second from Española, on the side towards India, where dwell a people who are considered by the neighboring islanders as most ferocious: and these feed upon human flesh. The same people have many kinds of canoes, in which they cross to all the surrounding islands and rob and plunder wherever they can; they are not different from other islanders, except that they wear their hair long, like women, and make use of the bows and javelins of cane,

* Probably Carib, the Indian name for Porto Rico.

with sharpened spear-points fixed on the thickest end, which I have before described, and therefore they are looked upon as ferocious, and regarded by the other Indians with unbounded fear; but I think no more of them than of the rest. These are the men who form unions with certain women who dwell alone in the island Matenin, which lies next to Española on the side towards India; these latter employ themselves in no labor suitable to their own sex, for they use bows and javelins as I have already described their paramours as doing, and for defensive armor have plates of brass, of which metal they possess great abundance. They assure me that there is another island larger than Española, whose inhabitants have no hair, and which abounds in gold more than any of the rest. I bring with me individuals of this island, and of the others that I have seen, who are proofs of the facts which I state. Finally, to compress into a few words the entire summary of my voyage and speedy return, and of the advantages derivable therefrom, I promise that, with a little assistance afforded me by our most invincible Sovereigns, I will procure them as much gold as they need, as great a quantity of spices, of cotton, and of mastic (which is only found in Chios), and as many men for the service of the navy as their Majesties may require. I promise, also, rhubarb and other sorts of drugs, which I am persuaded the men whom I have left in the aforesaid fortress have found already, and will continue to find; for I myself have tarried nowhere longer than I was compelled to do by the winds, except in the city of Navidad, while I provided for the building of the fortress, and took the necessary precautions for the perfect security of the men I left there. Although all I have related may appear to be wonderful and unheard of, yet the results of my voyage would have been more astonishing if I had had at my disposal such ships as I required. But these great and marvelous results are not to be attributed to any merit of mine, but to the holy Christian faith, and to the piety and religion of our Sovereigns, for that which the unaided intellect of man could not compass, the spirit of God has granted to human exertions, for God is wont to hear the prayers of his servants who love his precepts, even to the performance of apparent impossibilities. Thus it has happened to me in the present instance, who have accomplished a task to which the powers of mortal men had never hitherto attained; for if there have been those who have anywhere written or spoken of these islands, they have done so with doubts and conjectures, and no one has ever asserted that he has seen them, on which account their writings have been looked upon as little else than fables. Therefore let the King and Queen, our Princes and their most happy kingdoms, and all the other provinces of Christendom, render thanks to our Lord and Saviour Jesus Christ, who has granted us so great a victory and such prosperity. Let processions be made, and sacred feasts be held, and the temples be adorned with festive boughs. Let Christ rejoice on earth, as he rejoices in heaven, in the prospect of the salvation of the souls of so many nations hitherto lost. Let us also rejoice, as well on account of the exaltation of our faith as on account of the increase of our temporal prosperity, of which not only Spain, but all Christendom will be partakers.

Such are the events which I have briefly described. Farewell.

Lisbon,* the 14th of March.

CHRISTOPHER COLUMBUS,
Admiral of the Fleet of the Ocean.

* The date is one day after his leaving Lisbon, so it was probably written at sea.

LETTER TO LUIS DE SANTANGEL*

SENOR : Knowing the pleasure you will receive in hearing of the great victory which our Lord has granted me in my voyage, I hasten to inform you, that after a passage of seventy-one days, I arrived at the Indies, with the fleet of the most illustrious King and Queen, our Sovereigns, committed to my charge, where I discovered many islands, inhabited by people without number, and of which I took possession for their Highnesses by proclamation with the royal banner displayed, no one offering any contradiction. The first which I discov-

---

* This letter also narrates the discoveries of the first voyage, and largely repeats the information given in the letter to Sanchez, *ante*. Santangel was the secretary and steward of the household of Aragon, and had largely supplied the money for the expedition. The translation is by Samuel Kettell, printed in his *Personal Narrative of the First Voyage of Christopher Columbus*, . . . *Boston*, 1827. The original text is in Navarrete's *Coleccion de los Viages*, . . . *Madrid*, 1825.

ered I named San Salvador, in commemoration of our holy Saviour, who has, in a wonderful manner, granted all our success. The Indians call it Guanahani. To the second I gave the name of Santa Maria de Concepcion, to the third that of Fernandina, to the fourth that of Isabella, to the fifth that of Juana; thus giving each island a new name. I coasted along the island of Juana to the west, and found it of such extent that I took it for a continent, and imagined it must be the country of Cathay. Villages were seen near the sea-coast, but as I discovered no large cities, and could not obtain any communication with the inhabitants, who all fled at our approach, I continued on west, thinking I should not fail in the end to meet with great towns and cities; but having gone many leagues without such success, and finding that the coast carried me to the north, whither I disliked to proceed on account of the impending winter, I resolved to return to the south, and accordingly put about, and arrived at an excellent harbor in the island, where I dispatched two men into the country to ascertain whether the King or any large cities were in the neighborhood. They traveled three days, and met with innumerable settlements of the natives, of a small size, but did not succeed in finding any sovereign of the territory, and so returned. I made out to learn from some Indians, which I had before taken, that this was an island, and proceeded along the coast to the east an hundred and seven leagues, till I reached the extremity. I then discovered another island east of this, eighteen leagues distant, which I named Española, and followed its northern coast, as I did that of Juana, for the space of an hundred and seventy-eight leagues to the east. All these countries are of surpassing excellence, and in particular Juana, which contains abundance of fine harbors, excelling any in Christendom, as also many large and beautiful rivers. The land is high, and exhibits chains of tall mountains, which seem to reach to the skies, and surpass beyond comparison the isle of Cetrefrey. These display themselves in all manner of beautiful shapes. They are accessible in every part, and covered with a vast

variety of lofty trees, which it appears to me
never lose their foliage, as we found them fair
and verdant as in May in Spain. Some were
covered with blossoms, some with fruit, and
others in different stages, according to their
nature. The nightingale and a thousand other
sorts of birds were singing in the month of
November wherever I went. There are palm-
trees in these countries, of six or eight sorts,
which are surprising to see, on account of their
diversity from ours, but, indeed, this is the
case with respect to the other trees, as well as
the fruits and weeds. Beautiful forests of pines
are likewise found, and fields of vast extent.
Here are also honey, and fruits of thousand sorts,
and birds of every variety. The lands contain
mines of metals, and inhabitants without number. The island of Española is preëminent in
beauty and excellence, offering to the sight
the most enchanting view of mountains,
plains, rich fields for cultivation, and pastures
for flocks of all sorts, with situations for towns
and settlements. Its harbors are of such excellence that their description would not gain

belief, and the like may be said of its abundance of large and fine rivers, the most of which abound in gold. The trees, fruits, and plants of this island differ considerably from those of Juana, and the place contains a great deal of spicery, and extensive mines of gold and other metals. The people of this island, and of all the others which I have become acquainted with, go naked as they were born, although some of the women wear at the loins a leaf or bit of cotton cloth, which they prepare for that purpose. They do not possess iron, steel, or weapons, and seem to have no inclination for the latter, being timorous to the last degree. They have an instrument consisting of a cane, taken while in seed, and headed with a sharp stick, but they never venture to use it. Many times I have sent two or three men to one of their villages, when whole multitudes have taken to flight at the sight of them, and this was not by reason of any injury we ever wrought them, for at every place where I have made any stay, and obtained communication with them, I have made them presents of cloth and such other things as I possessed, without demanding anything in return. After they have shaken off their fear of us, they display a frankness and liberality in their behavior which no one would believe without witnessing it. No request of anything from them is ever refused, but they rather invite acceptance of what they possess, and manifest such a generosity that they would give away their own hearts. Let the article be of great or small value, they offer it readily, and receive anything which is tendered in return with perfect content. I forbade my men to purchase their goods with such worthless things as bits of platters and broken glass, or thongs of leather, although when they got possession of one of these, they estimated it as highly as the greatest jewel in the world. The sailors would buy of them for a scrap of leather pieces of gold weighing two castellanos and a half, and even more of this metal for something still less in value. The whole of an Indian's property might be purchased of him for a few blancas; this would amount to two or three castellanos' value of gold, or the same of cotton thread. Even the pieces of broken hoops from the casks they would receive in barter for their articles, with the greatest simplicity. I thought such traffic unjust, and therefore forbade it. I presented them with a variety of things, in order to secure their affection, and that they may become Christians, and enter into the services of their Highnesses and the Castilian nation, and also aid us in procuring such things as they possess, and we stand in need of. They are not idolaters, nor have they any sort of religion, except believing that power and goodness are in heaven, from which place they entertained a firm persuasion that I had come with my ships and men. On this account wherever we met them they showed us the greatest reverence after they had overcome their fear. Such conduct cannot be ascribed to their want of understanding, for they are a people of much ingenuity, and navigate all those seas, giving a remarkably good account of every part, but do not state that they have met with people in clothes or ships like ours. On my arrival at the Indies

I took by force from the first island I came to a few of the inhabitants, in order that they might learn our language, and assist us in our discoveries. We succeeded ere long in understanding one another by signs and words, and I have them now with me, still thinking we have come from heaven, as I learn by much conversation which I have had with them. This they were the first to proclaim wherever we went, and the other natives would run from house to house, and from village to village, crying out, "Come, and see the men from heaven!" so that all the inhabitants, both men and women, having gathered confidence, hastened toward us, bringing victuals and drink, which they presented to us with a surprising good will. In all the islands they possess a vast number of canoes, which are of various sizes, each one constructed of a single log, and shaped like a fusta. Some of these are as large as a fusta of eighteen oars, although narrow, on account of the material. I have seen sixty or eighty men in one of these canoes, and each man with his paddle. They are rowed with a swiftness which no boat can equal, and serve the purpose of transporting goods among these innumerable islands. I did not observe any great diversity in the appearance of the inhabitants in the different parts of these countries, nor in their customs nor language, for singularly enough in this last respect, they all understand one another; on which account I hope their Highnesses will exert themselves for the conversion of these people to our holy faith, in which undertaking they will be found very tractable. I have already related that I proceeded along the coast of Juana for an hundred and seven leagues from west to east, from which I dare affirm this island to be larger than England and Scotland together; for, besides the extent of it which I coasted, there are two unexplored provinces to the west, in one of which, called Cibau, are people with tails. These districts cannot be less than fifty or sixty leagues in extent, according as I learn from my Indians, who are acquainted with all these islands. The other island, called Española, is more extensive than the division of Spain from Corunna to Fontarabia, as I traversed one side of it for the distance of an hundred and thirty-eight leagues from west to east. This is a most beautiful island, and although I have taken possession of them all, in the name of their Highnesses, and every one remains in their power, and as much at their disposal as the kingdoms of Castile, and although they are all furnished with everything that can be desired, yet the preference must be given to Española, on account of the mines of gold which it possesses, and the facilities it offers for trade with continents and countries this side and beyond that of the Great Can, which traffic will be great and profitable. I have accordingly taken possession of a place, which I named Villa de Navidad, and built there a fortress which is at present complete, and furnished with a sufficiency of men for the enterprise; with these I have left arms, ammunitions, and provisions for more than a year, a boat, and expert men in all necessary arts. The King of the country has shown great friendship toward us, and held himself a brother to me. Even should their friendly inclinations change, and become hostile, yet nothing can be feared from them as they are totally ignorant in the world. The small number of men whom I have left there would be sufficient to ravage the whole territory, and they may remain there with perfect safety, taking proper care of themselves. In all the islands, as far as I could observe, the men are content with a single wife each, except that a chief or king has as many as twenty. The women appear to do more work than the men, and as to their property I have been unable to learn that they have any private possessions, but apparently all things are in common among them, especially provisions. In none of the islands hitherto visited have I found any people of monstrous appearance, according to the expectation of some, but the inhabitants are all of very pleasing aspect, not resembling the blacks of Guinea, as their hair is straight, and their color lighter. The rays of the sun are here very powerful, although the latitude is twenty-six degrees, but in the islands where there are high mountains the winter is cold, which the inhabitants endure from habit, and the use of hot spices with their food. An island situated in the second strait, at the entrance to the Indies, is peopled with inhabitants who eat live flesh, and are esteemed very ferocious in all the other parts. They possess many canoes with which they scour all the islands of India, robbing and capturing all they meet. They are not of a more deformed appearance than the others, except that they wear their hair long like women, and use bows and arrows, which last are made of cane and pointed with a stick for want of iron, which they do not possess. They exchange their wives, and although these are esteemed a fierce people among the neighboring islands, yet I do not regard them more than the others, as the most of the inhabitants of these regions are very great cowards. One of these islands is peopled solely by women, who practise no feminine occupations, but exercise the bow and arrow, and cover themselves with plates of copper, which metal they have in abundance. There is another island, as I am assured, larger than Española, in which the inhabitants are without hair, and which contains a great abundance of gold. In confirmation of these and other accounts I have brought the Indians along with me for testimonies. In conclusion, and to speak only of what I have performed: this voyage so hastily dispatched will, as their Highnesses may see, enable any desirable quantity of gold to be obtained by a very small assistance afforded me on their part. At present there are within reach: spices and cotton to as great an amount as they can desire; aloe, in a great abundance; and equal store of mastic, a production nowhere else found except in Greece and the island of Scio, where it is sold at such a price as the possessors choose. To these may be added slaves, as numerous as may be wished for. Besides, I have, as I think, discovered rhubarb and cinnamon, and expect countless other things of value will be found by the men whom I left there, as I made it a point not to stay in any one place while the wind enabled me to proceed upon the voyage, except at Villa de Navidad, where I left them well established. I should have accomplished much more, had those in the other vessels done their duty.* This is ever certain, that God grants to those that walk in his ways the performance of things which seem impossible, and this enterprise might in a signal manner have been considered so, for although many have talked of these countries, yet it has been nothing more than conjecture. Our Saviour having vouchsafed this victory to our most illustrious King and Queen and their kingdoms, famous for so eminent a deed, all Christendom should rejoice and give solemn thanks to the holy Trinity for the addition of as many people to our holy faith, and also for the temporal profit accruing not only to Spain, but to all Christians.

On board the caravel,† off the Azores, February 15th, 1493.

---

* This is a reference to the commanders of the *Pinta* and *Nina*, Pinzon and Yanez, who had proved unruly and insubordinate.

† The *Nina*. The ship had just outlived a furious storm, in which destruction seemed so imminent that Columbus and his crew drew lots as to who should make pilgrimages to the shrines of St. Mary of Guadalupe and St. Mary of

P. S. After writing the above, being at sea near Castile, the wind rose with such fury from the south and southeast that I was obliged to bear away, and run into the port of Lisbon,* where I escaped by the greatest miracle in the world. From this place I shall write to their Highnesses. Throughout the Indies I always found the weather like May. I made the passage thither in seventy-one days, and back in forty-eight, during thirteen of which number I was driven about by storms. The seamen here inform me that there was never known a winter in which so many ships were lost.†

March 4th.

Loretto; and, fearing the loss of the fact of his discovery, Columbus wrote an account of it on parchment, rolled it in waxed cloth, placed it in a wooden cask, and threw it into the sea.

* "When I was driven by a tempest into the port of Lisbon (having lost my sails), I was falsely accused (by those at court) of having put in thither with the intention of giving the Indies to the Sovereign of that country."

† Endorsed: "This letter Columbus sent to the steward of the household from the islands discovered in the Indies, in another to their Highnesses."

LETTER TO FERDINAND AND ISABELLA*

MOST HIGH AND MIGHTY SOVEREIGNS: In obedience to your Highnesses' commands, and with submission to superior judgment, I will say whatever occurs to me in reference to the colonization and commerce of the island of Española, and of the other islands, both those already discovered and those that may be discovered hereafter.

In the first place, as regards the island of Española: Inasmuch as the number of colonists who desire to go thither amounts to two

* This letter has been assigned to the year 1497, but the internal evidence indicates that it was written before Columbus sailed on his second voyage, as the number of colonists he speaks of as wishing to go agrees with the statements as to the size of the second expedition. This fixes the date between July 1st and September 25, 1493. It is thus the first suggestion of a code of American laws. The translation is by George Dexter, printed in the *Proceedings of the Massachusetts Historical Society*, Vol. XVI. The original text is in the *Cartas de Indias*, . . . *Madrid*, 1877.

thousand, owing to the land being safer and better for farming and trading, and because it will serve as a place to which they can return and from which they can carry on trade with the neighboring islands:

Item. That in the said island there shall be founded three or four towns, situated in the most convenient places, and that the settlers who are there be assigned to the aforesaid places and towns.

Item. That for the better and more speedy colonization of the said island, no one shall have liberty to collect gold in it except those who have taken out colonists' papers* and have built houses for their abode, in the town in which they are, that they may live united and in greater safety.

Item. That each town shall have its alcalde or alcaldes, and its notary public, as is the use and custom in Castile.

Item. That there shall be a church, and parish priests or friars to administer the sacra-

---

* Spanish: *tomaren veçindad.*

ments, to perform divine worship, and for the conversion of the Indians.

Item. That none of the colonists shall go to seek gold without a license from the governor or alcalde of the town where he lives; and that he must first take oath to return to the place whence he sets out, for the purpose of registering faithfully all the gold he may have found, and to return once a month, or once a week, as the time may have been set for him, to render account and show the quantity of said gold; and that this shall be written, done by the notary before the alcalde, or, if it seems better, that a friar or priest, deputed for the purpose, shall also be present.

Item. That the gold thus brought in shall be smelted immediately, and stamped with some mark that shall distinguish each town; and that the portion which belongs to your Highnesses shall be weighed, and given and consigned to each alcalde in his own town, and registered by the above-mentioned priest or friar, so that it shall not pass through the hands of only one person, and there shall be no opportunity to conceal the truth.

Item. That all gold that may be found without the mark of one of the said towns in the possession of any one who has once registered in accordance with the above order, shall be taken as forfeited, and that the accuser shall have one portion of it and your Highnesses the other.

Item. That one per centum of all the gold that may be found shall be set aside for building churches and adorning the same, and for the support of the priests or friars belonging to them; and, if it should be thought proper to pay any thing to the alcaldes or notaries for their services, or for ensuring the faithful performance of their duties, that this amount shall be sent to the governor or treasurer who may be appointed there by your Highnesses.

Item. As regards the division of the gold, and the share that ought to be reserved for your Highnesses, this, in my opinion, must be left to the aforesaid governor and treasurer, because it will have to be greater or less, according to the quantity of gold that may be found. Or, should it seem preferable, your Highnesses might, for the space of one year, take one-half, and the collector the other, and a better arrangement for the division be made afterward.

Item. That if the said alcaldes or notaries shall commit or be privy to any fraud, punishment shall be provided, and the same for the colonists who shall not have declared all the gold they have.

Item. That in the said island there shall be a treasurer, with a clerk to assist him, who shall receive all the gold belonging to your Highnesses, and the alcaldes and notaries of the towns shall each keep a record of what they deliver to the said treasurer.

Item. As, in the eagerness to get gold, every one will wish, naturally, to engage in its search in preference to any other employment, it seems to me that the privilege of going to look for gold ought to be withheld during some portion of each year, that there may be opportunity to have the other business necessary for the island performed.

Item. In regard to the discovery of new countries, I think permission should be granted to all that wish to go, and more liberality used in the matter of the fifth, making the tax easier, in some fair way, in order that many may be disposed to go on voyage.

I will now give my opinion about ships going to the said island of Española, and the order that should be maintained; and that is, that the said ships should only be allowed to discharge in one or two ports designated for the purpose, and should register there whatever cargo they bring or unload; and when the time for their departure comes, that they should sail from these same ports, and register all the cargo they take in, that nothing may be concealed.

Item. In reference to the transportation of gold from the island to Castile, that all of it should be taken on board the ship, both that belonging to your Highnesses and the property of every one else; that it should all be placed in one chest with two locks, with their keys, and that the master of the vessel keep one key and some person selected by the governor and treasurer the other; that there should come with the gold, for a testimony, a list of all that has been put into the said chest, properly marked, so that each owner may receive his own; and that, for the faithful performance of this duty, if any gold whatsoever is found outside of the said chest in any way, be it little or much, it shall be forfeited to your Highnesses.

Item. That all the ships that come from the said island shall be obliged to make their proper discharge in the port of Cadiz, and that no person shall disembark or other person be permitted to go on board until the ship has been visited by the person or persons deputed for that purpose, in the said city, by your Highnesses, to whom the master shall show all that he carries, and exhibit the manifest of all the cargo, that it may be seen and examined if the said ship brings any thing hidden and not known at the time of lading.

Item. That the chest in which the said gold has been carried shall be opened in the presence of the magistrates of the said city of Cadiz, and of the person deputed for that purpose by your Highnesses, and his own property be given to each owner. I beg your Highnesses to hold me in your protection; and I remain, praying our Lord God for your Highnesses' lives and the increase of much greater States,

.S.
S. A. S.
X M Y
$X\rho o$ FERENS.*

* A custom in Spain was to connect some pious supplication with signatures, and such unquestionably is the meaning of the prefatory initials that Columbus placed before the curious hybrid Greek-Latin letters which he used as a signature. Their exact meaning is doubtful, but the most probable explanation is, "*Salve me Xristus, Maria, Yosephus.*" See also in this connection, the Deed of Entail, *post.*

PRIVILEGES OF COLUMBUS*

This declares that what follows belongs, and should and ought to belong to the Admiral, Viceroy and Governor of the Indies, for the King and Queen, our Lords.

ACCORDING to the capitulation entered into with their Highnesses, and signed with their royal names, it appears very clearly that their Highnesses permit and grant to the said Admiral of the Indies, all the preëminences and prerogatives which the Admiral of Castile holds and enjoys; to whom, in right of

---

* Before sailing on his first voyage Columbus entered into stipulations with Ferdinand and Isabella (printed in *Kettell*), in which certain "recompense" in the shape of rights, titles, offices, and a percentage of all "goods, merchandise, pearls, precious stones, gold, silver, spices, and all other articles," were granted him. Disputes arose concerning the extent of these, and so this argument was prepared to prove his asserted rights. From two papers printed in *Memorials of Columbus* (pp. 67–72), it seems probable that it was written in 1497. A second and longer argument is printed in this volume, *post*. The translation is taken from *Memorials of Columbus*, . . . London, 1823. The original text is in *Codice Diplomatico Colombo-Americano*, . . . Genova, 1823.

his privilege, it is known that the third part of whatever he shall gain, belongs: and consequently the Admiral of the Indies is entitled to the third part of whatever he has acquired of the islands and mainland which he has discovered and may discover; and likewise he is to have the tenth and the eighth, as appears from the third and fifth article of the aforesaid capitulation.

And if it should be argued that the third part granted to the Admiral of Castile is to be understood as relating to movables, which he might acquire by sea; whereas the said islands being mainland, although acquired by sea, the third part of them cannot belong to the Admiral, in consequence of their being immovable:

To this the said Admiral replies, by saying that it is to be observed that, in the aforesaid capitulation, the said Admiral of Castile is nominated Admiral of the sea: and on that account the third part of whatever he may acquire by sea is granted to him, no jurisdiction nor office being granted to him in any other part whatsoever; and it would be very improper and unreasonable to grant him a part of what is not within his jurisdiction, it being a general maxim that *propter officium datur beneficium,* because the benefit has and ought to have a connection with the office, and not out of it. But the Admiral of the Indies was constituted and nominated, according to the tenor of the aforesaid capitulation, Admiral, not of the sea, but expressly of the Indies and of the mainland, which he has acquired in executing and discharging the said office of Admiral: and thus is to be understood and interpreted the privilege of the said Admiral of Castile, and the article which refers to it; it being sufficiently manifest that every thing is to be understood *secundum subjectam materiam, et secundum qualitatem personarum;* for by interpreting them otherwise, the said privilege and article would be of no utility to the aforesaid Admiral of the Indies; for if he does not take the third of the aforesaid Indies, of which he is Admiral, as he has not been constituted Admiral of the sea, he ought not even to take what he might gain by the sea, on account of its being out of his jurisdiction and office, so that the said article and constitution would be of no avail to him; and such a thing is not to be asserted, for whatever expression is introduced into a contract must have its full force, and not be regarded as superfluous: how much more so in a case like this, of so much importance, utility, and glory to their Highnesses, obtained at a very small expense, and without any peril to their honor, persons, or property, but with considerable peril, as was well known, to the life of, and not without heavy expense to, the Admiral? For which reason the tenth part only must be looked upon as very trifling (no mention being made of the eighth, because this belongs to him as his proportion), and so very small part for so great a service would be a recompense indeed inconsiderable. And the remark of the divine laws is here very apposite, *quia beneficia Principum sunt latissime interpretanda.* And, moreover, favors conferred by Princes ought to be understood in the most ample and complete sense; more especially by the most high and renowned Princes, such as their Highnesses are, from whom, more than from all other persons, the most ample favors are to be expected. And therefore the said third part, although it appears very small, belongs to the aforesaid Admiral: for we observe, in companies formed by merchants, that the industry and foresight of one partner are looked upon and held to be upon an equal footing with the money of another, and an equal share belongs to him of the gains resulting, although obtained by the money of the other: how much more ought this to be the case of the Admiral, who displayed astonishing and incredible industry, and was exposed to great labor and peril in his own person, brothers and family? Therefore, with so much the more reason he ought to have the third of all, as was really the intention of their Highnesses. And that such is truly the meaning, we see by this, that their Highnesses grant to such as go to the Indies five parts out of six, and to others, four parts out of five, and the administration of the land, without any peril, the road being now open, secure and made known to every body. And in confirmation of what I say, as is expressed in many privileges of the said Admiral of the Indies, the said Admiral went by command of their Highnesses to acquire, not ships, or vessels, or any other thing of the sea, but expressly islands and the mainland, as is specifically mentioned in the privilege (which might be more properly called a grace), at the beginning, where it is thus declared: "and because you, Christopher Columbus, go by our command to discover and acquire islands and the mainland," &c. Now if the whole acquisition was to be islands and mainland, it is a necessary consequence that the third part must be of what has been acquired; and being the third part of the acquisition, it is notorious that the third part of the islands and mainland acquired belongs to the said Admiral: and certainly there is no reason to doubt that, if in the beginning the aforesaid Admiral had demanded a greater part, it would have been granted to him, the whole of such acquisition being made by him, a thing of which nobody had any hope or expectation, and which was far beyond the knowledge and dominion of their Highnesses. This, then, is a complete and distinct answer to those who assert the contrary; and the third part of the said Indies and mainland justly and clearly appears to belong to the said Admiral.

That of the tenth is very clear. With respect to the eighth, although it be equally clear, I wish to observe:

If it be asserted against him, that he is not to have the said eighth of the merchandise and articles conveyed and exported in the vessels which went for discovery, to those which went to the pearl fishery, and to other parts of this Admiralty, whilst he remained in the island of Española upon the service of their Highnesses, because he contributed nothing toward their equipment; it is answered that the equipment of such vessels was not notified to him, nor was he called upon or informed of it at the time of their departure: and, therefore, as by law, to the ignorant, who can prove ignorance of any fact, no time elapses, but on the contrary such plea undoubtedly grants a legitimate excuse, and even complete restitution; therefore, in the actual case it should be understood and declared, that the Admiral performed his part by offering to contribute his part to the present: nor can he be blamed, but rather those who did not notify to him, what it was their duty to do, &c.

DEED OF ENTAIL*

IN the name of the most holy Trinity, who inspired me with the idea, and afterward made it perfectly clear to me, that I could navigate and go to the Indies from Spain, by traversing the ocean westwardly; which I communicated to the King, Don Ferdinand, and to the Queen, Doña Isabella, our sovereigns; and they were pleased to furnish me the necessary equipment of men and ships, and to make me their Admiral over the said ocean, in all parts lying to the west of an imaginary line, drawn from pole to pole, a hundred leagues west of the Cape de Verd and Azore Islands; also appointing me their Viceroy and Governor over all continents and islands that I

*Spanish: *Institution del Mayorazgos.* It was written February 22, 1498, before Columbus sailed on his third voyage. The translation is by Washington Irving, and is printed in his *Life and Voyages of Columbus*, . . . *New York*, 1828, where, by a curious error, it is given as the will of Columbus. The original text is in Navarrete's *Coleccion de los Viages*, . . . *Madrid*, 1825.

might discover beyond the said line westwardly; with the right of being succeeded in the said offices by my eldest son and his heirs for ever; and a grant of the tenth part of all things found in the said jurisdiction; and of all rents and revenues arising from it; and the eighth of all the lands and every thing else, together with the salary corresponding to my rank of Admiral, Viceroy, and Governor, and all other emoluments accruing thereto, as is more fully expressed in the title and agreement sanctioned by their Highnesses.

And it pleased the Lord Almighty, that in the year one thousand four hundred and ninety-two, I should discover the continent of the Indias and many islands, among them Española, which the Indians call Ayte, and the Monicongos, Cipango. I then returned to Castile to their Highnesses, who approved of my undertaking a second enterprise for farther discoveries and settlements; and the Lord gave me victory over the island of Española, which extends six hundred leagues, and I conquered it and made it tributary; and I discovered many islands inhabited by cannibals, and seven hundred to the west of Española, among which is Jamaica, which we call Santiago; and three hundred and thirty-three leagues of continent from south to west, besides a hundred and seven to the north, which I discovered in my first voyage; together with many islands, as may more clearly be seen by my letters, memorials, and maritime charts. And as we hope in God that before long a good and great revenue will be derived from the above islands and continent, of which, for the reasons aforesaid, belong to me the tenth and the eighth, with the salaries and emoluments specified above; and considering that we are mortal, and that it is proper for every one to settle his affairs, and to leave declared to his heirs and successors the property he possesses or may have a right to: Wherefore I have concluded to create an entailed estate (*mayorazgo*) out of the said eighth of the lands, places and revenues, in the manner which I now proceed to state:

In the first place, I am to be succeeded by

Don Diego, my son, who in case of death without children is to be succeeded by my other son, Ferdinand; and should God dispose of him also without leaving children, and without my having any other son, then my brother, Don Bartholomew, is to succeed; and after him his eldest son; and if God should dispose of him without heirs, he shall be succeeded by his sons from one to another for ever; or, in the failure of a son, to be succeeded by Don Ferdinand, after the same manner, from son to son successively; or in their place by my brothers Bartholomew and Diego. And should it please the Lord that the estate, after having continued some time in the line of any of the above successors, should stand in need of an immediate and lawful male heir, the succession shall then devolve to the nearest relation, being a man of legitimate birth, and bearing the name of Columbus derived from his father and his ancestors. This entailed estate shall in nowise be inherited by a woman, except in case that no male is to be found, either in this or any other quarter of the world, of my real lineage, whose name, as well as that of his ancestors, shall have always been Columbus. In such an event (which may God forefend), then the female of legitimate birth most nearly related to the preceding possessor of the estate, shall succeed to it; and this is to be under the conditions herein stipulated at foot, which must be understood to extend as well to Don Diego, my son, as to the aforesaid and their heirs, every one of them, to be fulfilled by them; and failing to do so they are to be deprived of the succession for not having complied with what shall herein be expressed; and the estate to pass to the person most nearly related to the one who held the right: and the person thus succeeding shall in like manner forfeit the estate, should he also fail to comply with said conditions; and another person, the nearest of my lineage, shall succeed, provided he abide by them, so that they may be observed for ever in the form prescribed. This forfeiture is not to be incurred for trifling matters, originating in lawsuits, but in important cases, when the glory of God, or my own, or that of my family, may be concerned, which supposes a perfect fulfilment of all the things hereby ordained; all which I recommend to the courts of justice. And I supplicate his Holiness, who now is, and those that may succeed in the holy Church, that if it should happen that this, my will and testament, has need of his holy order and command for its fulfilment, that such order be issued in virtue of obedience, and under penalty of excommunication, and that it shall not be in any wise disfigured. And I also pray the King and Queen, our sovereigns, and their eldest-born, Prince Don Juan, our lord, and their successors, for the sake of the services I have done them, and because it is just, and that it may please them not to permit this my will and constitution of my entailed estate to be any way altered, but to leave it in the form and manner which I have ordained, for ever, for the greater glory of the Almighty, and that it may be the root and basis of my lineage, and a memento of the services I have rendered their Highnesses; that, being born in Genoa, I came over to serve them in Castile, and discovered, to the west of terra firma, the Indias and islands before mentioned. I accordingly pray their Highnesses to order that this, my privilege and testament, be held valid, and be executed summarily, and without any opposition or demur, according to the letter. I also pray the grandees of the realm, and the lords of the council, and all others having administration of justice, to be pleased not to suffer this my will and testament to be of no avail, but to cause it to be fulfilled as by me ordained; it being just that a noble, who has served the King and Queen, and the kingdom, should be respected in the disposition of his estate by will, testament, institution of entail or inheritance, and that the same be not infringed either in whole or in part.

In the first place, my son Don Diego, and all my successors and descendants, as well as my brothers Bartholomew and Diego, shall bear my arms, such as I shall leave them after my days, without inserting any thing else in them; and they shall be their seal to seal withal.

Don Diego my son, or any other who may inherit this estate, on coming into possession of the inheritance, shall sign with the signature which I now make use of, which is an X with an S over it, and an M with a Roman A over it, and over that an S, and then a Greek Y, with an S over it, with its lines and points as is my custom, as may be seen by my signatures, of which there are many, and it will be seen by the present one.

He shall only write "the Admiral," whatever other titles the King may have conferred on him. This is to be understood as respects his signature, but not the enumeration of his titles, which he can make at full length if agreeable, only the signature is to be "the Admiral."

The said Don Diego, or any other inheritor of this estate, shall possess my offices of Admiral of the ocean, which is to the west of an imaginary line, which his Highness ordered to be drawn, running from pole to pole a hundred leagues beyond the Azores, and as many more beyond the Cape de Verd Islands, over all which I was made, by their order, their Admiral of the sea, with all the preëminences held by Don Henrique in the Admiralty of Castile, and they made me their Governor and Viceroy perpetually and for ever, over all the islands and mainland discovered, or to be discovered, for myself and heirs, as is more fully shown by my treaty and privilege as above mentioned.

Item. The said Don Diego, or any other inheritor of this estate, shall distribute the revenue which it may please our Lord to grant him, in the following manner, under the above penalty:

First—Of the whole income of this estate, now and at all times, and of whatever may be had or collected from it, he shall give the fourth part of it to my brother, Don Bartholomew Columbus, Adelantado of the Indies; and this is to continue till he shall have acquired an income of a million of maravedises* for his support, and for the services he has rendered and will continue to render to this entailed estate;

---

* Approximately thirty-five hundred dollars, equivalent at the time to between ten and twelve thousand dollars.

which million he is to receive, as stated, every year, if the said fourth amount to so much, and that he have nothing else; but if he possess a part or the whole of that amount in rents, that thenceforth he shall not enjoy the said million, nor any part of it, except that he shall have in the said fourth part unto the said quantity of a million, if it should amount to so much; and as much as he shall have a revenue besides this fourth part, whatever sum of maravedises of known rent from property or perpetual offices, the said quantity of rent or revenue from property or offices shall be discounted; and from the said million shall be reserved whatever marriage-portion he may receive with any female he may espouse; so that whatever he may receive in marriage with his wife, no deduction shall be made on that account from said million, but only for whatever he may acquire or may have over and above his wife's dowry; and when it shall please God that he or his heirs and descendants shall derive from their property and offices a revenue of a million arising from rents, neither he nor his heirs shall enjoy any longer anything from the said fourth part of the entailed estate, which shall remain with Don Diego, or whoever may inherit it.

Item. From the revenues of the said estate, or from any other fourth part of it (should its amount be adequate to it), shall be paid every year to my son Ferdinand two millions, till such time as his revenue shall amount to two millions, in the same form and manner as in the case of Bartholomew, who, as well as his heirs, are to have the million or the part that may be wanting.

Item. The said Don Diego or Don Bartholomew shall make out of the said estate, for my brother Diego, such provision as may enable him to live decently, as he is my brother, to whom I assign no particular sum, as he has attached himself to the Church, and that will be given him which is right; and this to be given him in a mass, and before anything shall have been received by Ferdinand my son, or Bartholomew my brother, or their heirs, and also according to the amount of the income of the estate. And in case of discord, the case is to be referred to two of our relations, or other men of honor; and should they disagree among themselves, they will choose a third person as arbitrator, being virtuous and not distrusted by either party.

Item. All this revenue which I bequeath to Bartholomew, to Ferdinand, and to Diego, shall be delivered to and received by them as prescribed under the obligation of being faithful and loyal to Diego my son, or his heirs, they as well as their children; and should it appear that they, or any of them had proceeded against him in anything touching his honor, or the prosperity of the family, or of the estate, either in word or deed, whereby might come a scandal and debasement to my family, and a detriment to my estate; in that case, nothing farther shall be given to them or him, from that time forward, inasmuch as they are always to be faithful to Diego and to his successors.

Item. As it was my intention, when I first instituted this entailed estate, to dispose, or that my son Diego should dispose for me, of the tenth part of the income in favor of necessitous persons, as a tithe, and in commemoration of the almighty and eternal God; and persisting still in this opinion, and hoping that his High Majesty will assist me, and those who may inherit it, in this or the new world, I have resolved that the said tithe shall be paid in the manner following:

First—It is to be understood that the fourth part of the revenue of the estate which I have ordained and directed to be given to Don Bartholomew, till he have an income of one million, includes the tenth of the whole revenue of the estate; and that as in proportion as the income of my brother Don Bartholomew shall increase, as it has to be discounted from the revenue of the fourth part of the entailed estate, that the said revenue shall be calculated, to know how much the tenth part amounts to; and the part which exceeds what is necessary to make up the million for Don Bartholomew shall be received by such of my family as may most stand in need of it, discounting it from the said tenth, if their income do not amount to fifty thousand maravedises; and should any of these come to have an income to this amount, such a part shall be awarded them as two persons, chosen for the purpose, may determine, along with Don Diego or his heirs. Thus, it is to be understood that the million which I leave to Don Bartholomew comprehends the tenth of the whole revenue of the estate; which revenue is to be distributed among my nearest and most needy relations in the manner I have directed; and when Don Bartholomew has an income of one million, and that nothing more shall be due to him on account of said fourth part, then Don Diego, my son, or the persons which I shall herein point out, shall inspect the accounts, and so direct that the tenth of the revenue shall still continue to be paid to the most necessitous members of my family that may be found in this or any other quarter of the world, who shall diligently be sought out; and they are to be paid out of the fourth part from which Don Bartholomew is to derive his million; which sums are to be taken into account, and deducted from the said tenth, which, should it amount to more, the overplus, as it arises from the fourth part, shall be given to the most necessitous persons as aforesaid; and, should it not be sufficient, that Don Bartholomew shall have it until his own estate goes on increasing, leaving the said million in part or in the whole.

Item. The said Don Diego my son, or whoever may be the inheritor, shall appoint two persons of conscience and authority, and most nearly related to the famity, who are to examine the revenue and its amount carefully, and to cause the said tenth to be paid out of the fourth from which Don Bartholomew is to receive his million, to the most necessitated members of my family that may be found here or elsewhere, whom they shall look for diligently upon their consciences; and as it might happen that said Don Diego, or others after him, for reasons which may concern their own welfare, or the credit and support of the estate, may be unwilling to make known the full amount of the income; nevertheless I charge him on his conscience to pay the sum aforesaid; and I charge them on their souls and consciences not to denounce or make it known, except with the consent of Don Diego, or the person that may succeed him; but let the above tithe be paid in the manner I have directed.

Item. In order to avoid all disputes in the choice of the two nearest relations who are to act with Don Diego or his heirs, I hereby elect Don Bartholomew, my brother, for one, and Don Fernando, my son, for the other; and when these two shall enter upon the business they shall choose two other persons among the most trusty, and most nearly related, and these again shall elect two others when it shall be question of commencing the examination; and thus it shall be managed with diligence from one to the other, as well in this as in the other of government, for the service and glory of God, and the benefit of the said entailed estate.

Item. I also enjoin Diego, or any one that may inherit the estate, to have and maintain in the city of Genoa one person of our lineage to reside there with his wife, and appoint him a sufficient revenue to enable him to live decently, as a person closely connected with the family, of which he is to be the root and basis in that city; from which great good may accrue to him, inasmuch as I was born there, and came from thence.

Item. The said Don Diego, or whoever shall inherit the estate, must remit in bills, or in any other way, all such sums as he may be able to save out of the revenue of the estate, and direct purchases to be made in his name, or that of his heirs, in a fund in the Bank of St. George,* which gives an interest of six per cent. and is secure money; and this shall be devoted to the purpose I am about to explain.

Item. As it becomes every man of rank and property to serve God, either personally or by means of his wealth, and as all moneys deposited with St. George are quite safe, and Genoa is a noble city, and powerful by sea, and as at the time that I undertook to set out upon that discovery of the Indias, it was with the intention of supplicating the King and Queen, our lords, that whatever moneys should be derived from the said Indias should be invested in the conquest of Jerusalem; and as I did so suppli-

* The great financial corporation of Genoa.

cate them; if they do this, it will be well; if not, at all events the said Diego, or such person as may succeed him in this trust, to collect together all the money he can, and accompany the King our lord, should he go to the conquest of Jerusalem, or else go there himself with all the force he can command; and in pursuing this intention, it will please the Lord to assist toward the accomplishment of the plan; and should he not be able to effect the conquest of the whole, no doubt he will achieve it in part. Let him therefore collect and make a fund of all his wealth in St. George of Genoa, and let it multiply there till such time as it may appear to him that something of consequence may be effected as respects the project on Jerusalem; for I believe that when their Highnesses shall see that this is contemplated, they will wish to realize it themselves, or will afford him, as their servant and vassal, the means of doing it for them.

Item. I charge my son Diego and my descendants, especially whoever may inherit this estate, which consists, as aforesaid, of the tenth of whatsoever may be had or found in the Indias, and the eighth part of the lands and rents, all which, together with my rights and emoluments as Admiral, Viceroy and Governor, amount to more than twenty-five per cent—I say, that I require of him to employ all this revenue, as well as his person and all the means in his power, in well and faithfully serving and supporting their Highnesses, or their successors, even to the loss of life and property; since it was their Highnesses, next to God, who first gave me the means of getting and achieving this property, although, it is true, I came over these realms to invite them to the enterprise, and that a long time elapsed before any provision was made for carrying it into execution; which, however, is not surprising, as this was an undertaking of which all the world was ignorant, and no one had any faith in it; wherefore I am by so much the more indebted to them, as well as because they have since also much favored and promoted me.

Item. I also require of Diego, or whosoever may be in possession of the estate, that in the case of any schism taking place in the

Church of God, or that any person of whatever class or condition should attempt to despoil it of its property and honors, they hasten to offer at the feet of his holiness, that is, if they are not heretics (which God forbid), their persons, power and wealth, for the purpose of suppressing such schism, and preventing any spoliation of the honor and property of the Church.

Item. I command the said Diego, or whoever may possess the said estate, to labor and strive for the honor, welfare and aggrandizement of the city of Genoa, and to make use of all his power and means in defending and enhancing the good and credit of that republic, in all things not contrary to the service of the Church of God, or the high dignity of the King and Queen, our lords, and their successors.

Item. The said Diego, or whoever may possess or succeed to the estate, out of the fourth part of the whole revenue, from which, as aforesaid, is to be taken the tenth, when Don Bartholomew or his heirs shall have saved the two millions, or part of them, and when the time shall come of making a distribution among our relations, shall apply and invest the said tenth in providing marriages for such daughters of our lineage as may require it, and in doing all the good in their power.

Item. When a suitable time shall arrive, he shall order a church to be built in the island of Española, and in the most convenient spot, to be called Santa Maria de la Concepcion; to which is to be annexed an hospital, upon the best possible plan, like those of Italy and Castile, and a chapel be erected to say mass in for the good of my soul, and those of my ancestors and successors with great devotion, since no doubt it will please the Lord to give us a sufficient revenue for this and the aforementioned purposes.

Item. I also order Diego my son, or whosoever may inherit after him, to spare no pains in having and maintaining in the island of Española, four good professors of theology, to the end and aim of their studying and laboring to convert to our holy faith the inhabitants of the Indies; and in proportion as, by God's will the revenue of the estate shall increase, in the same degree shall the number of teachers and devout persons increase, who are to strive to make Christians of the natives; in attaining which no expense should be thought too great. And in commemoration of all that I hereby ordain, and of the foregoing, a monument of marble shall be erected in the said Church of la Concepcion, in the most conspicuous place, to serve as a record of what I here enjoin on the said Diego, as well as to other persons who may look upon it; which marble shall contain an inscription to the same effect.

Item. I also require of Diego my son, and whosoever may succeed him in the estate, that every time, and as often as he confesses, he first show his obligation, or a copy of it, to the confessor, praying him to read it through, that he may be enabled to inquire respecting its fulfillment; from which will redound great good and happiness to his soul.

Seville, February 22, 1498.

                        S.
                  S   A   S
                  X   M   Y
            EL ALMIRANTE.

LETTER TO FERDINAND AND ISABELLA*

MOST serene and most exalted and powerful Princes, the King and Queen, our Sovereigns: The Blessed Trinity moved your Highnesses to the encouragement of this enterprise to the Indies, and of its infinite goodness has made me your messenger therein; as ambassador for which undertaking I approached your royal presence, moved by the consideration that I was appealing to the most exalted monarchs in Christendom, who exercised so great an influence over the Christian faith and its advancement in the world; those who heard of it looked upon it as impossible, for they fixed all

---

* Describing his third voyage to America, for which he sailed May 30, 1498. He reached San Domingo on August 30th of the same year, and must have written this account either immediately before or after that date, and despatched it to Spain. The translation is by R. H. Major, and is printed in his *Select Letters of Christopher Columbus*, . . . London, 1848. The original text is in Navarrete's *Coleccion de los Viages*, . . . *Madrid*, 1825.

their hopes on the favors of fortune, and pinned their faith solely upon chance. I gave to the subject six or seven years of great anxiety,* explaining, to the best of my ability, how great service might be done to our Lord by this undertaking, in promulgating His sacred name and our holy faith among so many nations—an enterprise so exalted in itself, and so calculated to enhance the glory and immortalize the renown of the greatest sovereigns. It was also requisite to refer to the temporal prosperity which was foretold in the writings of so many trustworthy and wise historians, who related that great riches were to be found in these parts. And at the same time I thought it desirable to bring to bear upon the subject the sayings and opinions of those who have written upon the geography of the world. And, finally, your Highnesses came to the determination that the undertaking should be entered upon. In this your Highnesses exhibited the noble spirit which has been always manifested

---

* Columbus first applied to Spain in 1485.

by you on every subject; for all others who had thought of the matter, or heard it spoken of, unanimously treated it with contempt, with the exception of two friars,* who always remained constant in their belief of its practicability. I, myself, in spite of fatiguing opposition, felt sure that the enterprise would, nevertheless, prosper, and continue equally confident of it to this day, because it is a truth that, though everything will pass away, the Word of God will not; and I believe that every prospect which I hold out will be accomplished; for it was clearly predicted concerning these lands, by the mouth of the prophet Isaiah, in many places in Scripture, that from Spain the holy name of God was to be spread abroad. Thus I departed in the name of the Holy Trinity, and returned very soon, bringing with me an account of the practical fulfilment of everything I had said. Your Highnesses again sent me out, and in a short

* Fray Juan Perez de Marchena, a Franciscan, keeper of the Convent of de la Rabida, and Fray Diego de Deza, a Dominican.

space of time, by God's mercy, not by* I discovered three hundred and thirty-three leagues of terra firma on the eastern side,† and seven hundred islands, besides those which I discovered on the first voyage; I also succeeded in circumnavigating the island of Española, which is larger in circumference than all Spain, the inhabitants of which are countless, and all of whom may be laid under tribute. It was then that complaints arose, disparaging the enterprise that I had undertaken, because, forsooth, I had not immediately sent the ships home laden with gold—no allowance being made for the shortness of the time, and all the other impediments of which I have already spoken. On this account (either as a punishment of my sins, or, as I trust, for my salvation) I was held in detestation, and had obstacles placed in the way of everything I said, or for which I petitioned. I therefore resolved to apply to your Highnesses, to inform you of all the won-

* A blank occurs in the MS.
† Columbus here confused the island of Cuba with the main coast, which he did not see till the present voyage.

derful events that I had experienced, and to explain the reason of every proposition that I made, making reference to the nations that I had seen, among whom, and by whose instrumentality, many souls may be saved. I related how the natives of Española had been laid under tribute to your Highnesses, and regarded you as their sovereigns. And I laid before your Highnesses abundant samples of gold and copper—proving the existence of extensive mines of those metals. I also laid before your Highnesses many sorts of spices, too numerous to detail; and I spoke of the great quantity of brazil-wood, and numberless other articles found in those lands. All this was of no avail with some persons, who began, with determined hatred, to speak ill of the enterprise, not taking into account the service done to our Lord in the salvation of so many souls, nor the enhancement of your Highnesses' greatness to a higher pitch than any earthly prince has yet enjoyed; nor considering that, from the exercise of your Highnesses' goodness, and the expense incurred, both spiritual and temporal advantage was to be expected, and that Spain must in the process of time derive from thence, beyond all doubt, an unspeakable increase of wealth. This might be manifestly seen by the evidences already given in writing in the descriptions of the voyages already made, which also prove that the fulfilment of every other hope may be reasonably expected. Nor were they affected by the consideration of what great princes throughout the world have done to increase their fame: as, for example, Solomon, who sent from Jerusalem to the uttermost parts of the east, to see Mount Sopora, in which expedition his ships were detained three years; and which mountain your Highnesses now possess in the island of Española. Nor, as in the case of Alexander, who sent to observe the mode of government in the island of Taprobana, in India; and Cæsar Nero, to explore the sources of the Nile, and to learn the causes of its increase in the spring, when water is needed; and many other mighty deeds that princes have done, and which it is allotted to princes to achieve. Nor was it of any avail that no prince of Spain, as far as I have read, has ever hitherto gained possession of land out of Spain; and that the world of which I speak is different from that in which the Romans, and Alexander, and the Greeks made mighty efforts with great armies to gain the possession of. Nor have they been affected by the recent noble example of the kings of Portugal, who have had the courage to explore as far as Guinea, and to make the discovery of it, expending so much gold and so many lives in the undertaking that a calculation of the population of the kingdom would show that one-half of them have died in Guinea: and though it is now a long time since they commenced these great exertions, the return for their labor and expense has hitherto been but trifling; this people has also dared to make conquests in Africa, and to carry on their exploits to Ceuta, Tangier, and Alcazar, repeatedly giving battle to the Moors, and all this at great expense, simply because it was an exploit worthy of a prince, undertaken for the service of God and to advance the enlargement of His kingdom.

The more I said on the subject the more twofold was reproach cast upon it, even to the expression of abhorrence, no consideration being given to the honor and fame that accrued to your Highnesses throughout all Christendom, from your Highnesses' having undertaken this enterprise; so that there was neither great nor small who did not desire to hear tidings of it. Your Highnesses replied to me encouragingly, and desired that I should pay no regard to those who spoke ill of the undertaking, inasmuch as they had received no authority or countenance whatever from your Highnesses.

I started from San Lucar, in the name of the most Holy Trinity, on Wednesday, the 30th of May, much fatigued with my voyage, for I had hoped, when I left the Indies, to find repose in Spain; whereas, on the contrary, I experienced nothing but opposition and vexation. I sailed to the island of Madeira by a circuitous route, in order to avoid any encounter with an armed fleet from France, which was on the lookout for me off Cape St. Vincent. Thence I went

to the Canaries, from which islands I sailed
with but one ship and two caravels, having dispatched the other ships to Española by the direct road to the Indies; while I myself moved
southward, with the view of reaching the equinoctial line, and of then proceeding westward,
so as to leave the island of Española to the
north. But having reached the Cape Verd
Islands (an incorrect name, for they are so bar-
ren that nothing green was to be seen there,
and the people so sickly that I did not venture
to remain among them), I sailed away four
hundred and eighty miles, which is equivalent
to a hundred and twenty leagues, toward the
southwest, where, when it grew dark, I found
the north star to be in the fifth degree. The wind
then failed me, and I entered a climate where
the intensity of the heat was such that I thought
both ships and men would have been burned up,
and everything suddenly got into such a state
of confusion that no man dared go below deck
to attend to the securing of the water-cask and
the provisions. This heat lasted eight days;
on the first day the weather was fine, but on

the seven other days it rained and was cloudy, yet we found no alleviation of our distress; so that I certainly believe that if the sun had shone as on the first day, we should not have been able to escape in any way.

I recollect that, in sailing toward the Indies, as soon as I passed a hundred leagues to the westward of the Azores, I found the temperature change: and this is so all along from north to south. I determined, therefore, if it should please the Lord to give me a favorable wind and good weather, so that I might leave the part where I then was, that I would give up pursuing the southward course, yet not turn backward, but sail toward the west, moving in that direction in the hope of finding the same temperature that I had experienced when I sailed in the parallel of the Canaries, and then, if it proved so, I should still be able to proceed more to the south. At the end of these eight days it pleased our Lord to give me a favorable east wind, and I steered to the west, but did not venture to move lower down toward the south, because I discovered a very great change in the sky and the stars, although I found no alteration in the temperature. I resolved, therefore, to keep on the direct westward course, in a line from Sierra Leone, and not to change on another tack, which I was very desirous to do, for the purpose of repairing the vessels, and of renewing, if possible, our stock of provisions, and taking in what water we wanted. At the end of seventeen days, during which our Lord gave me a propitious wind, we saw land at noon, of Tuesday, the 31st of July. This I had expected on the Monday before, and held that route up to this point; but as the sun's strength increased, and our supply of water was failing, I resolved to make for the Caribee Islands, and set sail in that direction; when, by the mercy of God, which He has always extended to me, one of the sailors* went up to the maintop, and saw to the westward a range of three mountains. Upon this we repeated the "Salve

---

*Navarrete states that this was a servant of Columbus named Alonzo Perez.

Regina," and other prayers, and all of us gave many thanks to our Lord. I then gave up our northward course, and put in for the land: at the hour of complines we reached a cape, which I called Cape Galea,* having already given to the island the name of Trinidad, and here we found a harbor, which would have been excellent, but there was no good anchorage. We saw houses and people on the spot, and the country around was very beautiful, and as fresh and green as the gardens of Valencia in the month of March. I was disappointed at not being able to put into the harbor, and ran along the coast to the westward. After sailing five leagues I found very good bottom, and anchored. The next day I set sail in the same direction, in search of a harbor where I might repair the vessels and take in water, as well as improve the stock of provisions which I had brought out with me. When we had taken in a pipe of water, we pro-

* Cape Galeota, the southeastern point of Trinidad. It was now that Columbus first saw the main coast of America.

ceeded onward till we reached the cape, and there, finding good anchorage and protection from the east wind, I ordered the anchors to be dropped, the water-cask to be repaired, a supply of water and wood to be taken in, and the people to rest themselves from the fatigues which they had endured for so long a time. I gave to this point the name of Sandy Point (*Punta del Arenal*). All the ground in the neighborhood was filled with foot-marks of animals, like the impression of the foot of a goat; but although it would have appeared from this circumstance that they were very numerous, only one was seen, and that was dead. On the following day a large canoe came from the eastward, containing twenty-four men, all in the prime of life, and well provided with arms, such as bows, arrows, and wooden shields; they were all, as I have said, young, well-proportioned, and not dark black, but whiter than any other Indians that I had seen, of very graceful gesture, and handsome forms, wearing their hair long and straight, and cut in the Spanish style. Their heads were bound round with cotton scarfs elaborately worked in colors, which resembled the Moorish head-dresses. Some of these scarfs were worn round the body and used as a covering in lieu of trousers. The natives spoke to us from the canoe while it was yet at a considerable distance, but none of us could understand them; I made signs to them, however, to come nearer to us, and more than two hours were spent in this manner; but if by any chance they moved a little nearer, they soon pushed off again. I caused basins and other shining objects to be shown to them to tempt them to come near; and after a long time they came somewhat nearer than they had hitherto done, upon which, as I was anxious to speak with them, and had nothing else to show them to induce them to approach, I ordered a drum to be played upon the quarter-deck, and some of our young men to dance, believing the Indians would come to see the amusement. No sooner, however, did they perceive the beating of the drum and the dancing, than they all left their oars, and strung their bows, and each man laying hold of his shield, they commenced discharging their arrows at us; upon this, the music and dancing soon ceased, and I ordered a charge to be made from some of our crossbows; they then left us, and went rapidly to the other caravel, and placed themselves under its poop. The pilot of that vessel received them courteously, and gave to the man who appeared to be their chief a coat and hat; and it was then arranged between them that he should go to speak with him on shore. Upon this the Indians immediately went thither and waited for him; but as he would not go without my permission, he came to my ship in the boat, whereupon the Indians got into their canoe again and went away, and I never saw any more of them or of any of the other inhabitants of the island. When I reached the point of Arenal, I found that the island of Trinidad formed with the land of Gracia a strait of two leagues' width from east to west,* and as we had to pass through it to go to the north, we found

* This was the Gulf of Paria, and the currents were occasioned by the river Orinoco, which empties into it.

some strong currents which crossed the strait, and which made a great roaring, so that I concluded there must be a reef of sand or rocks, which would preclude our entrance; and behind this current was another and another, all making a roaring noise like the sound of breakers against the rocks. I anchored there, under the said point of Arenal, outside of the strait, and found the water rush from east to west with as much impetuosity as that of the Guadalquivir at its conflux with the sea; and this continued constantly day and night, so that it appeared to be impossible to move backward for the current or forward for the shoals. In the dead of night, while I was on deck, I heard an awful roaring that came from the south, toward the ship; I stopped to observe what it might be, and I saw the sea rolling from west to east like a mountain, as high as the ship, and approaching by little and little; on the top of this rolling sea came a mighty wave roaring with a frightful noise, and with all this terrific uproar were other conflicting currents, producing, as I have already

said, a sound as of breakers upon the rocks.
To this day I have a vivid recollection of the
dread I then felt, lest the ship might founder
under the force of that tremendous sea; but it
passed by, and reached the mouth of the before-
mentioned passage, where the uproar lasted
for a considerable time. On the following day
I sent out boats to take soundings, and found
that in the strait, at the deepest part of the
embouchure, there were six or seven fathoms
of water, and that there were constant contrary
currents, one running inward, and the other
outward. It pleased the Lord, however, to
give us a favorable wind, and I passed through
the middle of the strait, after which I recovered
my tranquillity. The men happened at this
time to draw up some water from the sea,
which, strange to say, proved to be fresh. I
then sailed northward till I came to a very
high mountain, at about twenty-six leagues
from the Punta del Arenal; here two lofty
headlands appeared, one toward the east, and
forming part of the island of Trinidad, and the
other, on the west, being part of the land

which I have already called Gracia; we found here a channel still narrower than that of Arenal, with similar currents, and a tremendous roaring of water; the water here also was fresh. Hitherto I had held no communication with any of the people of this country, although I very earnestly desired it; I therefore sailed along the coast westward, and the further I advanced, the fresher and more wholesome I found the water; and when I had proceeded a considerable distance, I reached a spot where the land appeared to be cultivated. There I anchored, and sent the boats ashore, and the men who went in them found the natives had already left the place; they also observed that the mountain was covered with monkeys. They came back, and as the coast at that part presented nothing but a chain of mountains, I concluded that further west we should find the land flatter, and consequently in all probability inhabited. Actuated by this thought I weighed anchor, and ran along the coast until we came to the end of the cordillera; I then anchored at the mouth of a river, and we were soon visited by a great number of the inhabitants, who informed us that the country was called Paria, and that further westward it was more fully peopled. I took four of these natives, and proceeded on my westward voyage, and when I had gone eight leagues further, I found on the other side of a point which I called the Needle, one of the most lovely countries in the world, and very thickly peopled; it was 3 o'clock in the morning when I reached it, and seeing its verdure and beauty, I resolved to anchor there and communicate with the inhabitants. Some of the natives soon came out to the ship, in canoes, to beg me, in the name of their King, to go on shore; and when they saw that I paid no attention to them, they came to the ship in their canoes in countless numbers, many of them wearing pieces of gold on their breasts, and some with bracelets of pearls on their arms; on seeing which I was much delighted, and made many inquiries, with the view of learning where they found them. They informed me that they were to be procured in their own neighborhood, and also at a spot to the northward of that place. I would have remained here, but the provisions of corn, and wine, and meats, which I had brought out with so much care for the people whom I had left behind, were nearly wasted, so that all my anxiety was to get them into a place of safety, and not to stop for any thing. I wished, however, to get some of the pearls that I had seen, and with that view sent the boats on shore. The natives are very numerous, and for the most part handsome in person, and of the same color as the Indians we had already seen; they are, moreover, very tractable, and received our men who went on shore most courteously, seeming very well disposed toward us. These men relate, that when the boats reached shore, two of the chiefs, whom they took to be father and son, came forward in advance of the mass of the people, and conducted them to a very large house with façades, and not round and tent-shaped as the other houses were; in this house were many seats, on which they made our men sit down, they themselves sitting on other seats. They then caused bread to be brought, with many kinds of fruits, and various sorts of wine, both white and red, not made of grapes, but apparently produced from different fruits. The most reasonable inference is, that they use maize, which is a plant that bears a spine like an ear of wheat, some of which I took with me from Spain, where it grows abundantly; this they seem to regard as most excellent, and set a great value upon it. The men remained together at one end of the house, and the women at the other. Great vexation was felt by both parties that they could not understand each other, for they were mutually anxious to make inquiries respecting each other's country. After our men had been entertained at the house of the elder Indian, the younger took them to his house, and gave them an equally cordial reception, after which they returned to their boats and came on board. I weighed anchor forthwith, for I was hastened by my anxiety to save the provisions, which were becoming spoiled, and which I had procured and preserved with so much care and trouble, as well as to attend to my own health, which had been affected by long watching; and although on my former voyage, when I discovered terra firma, I passed thirty-three days without natural rest, and was all that time deprived of sight, yet never were my eyes so much affected or so painful as at this period. These people, as I have already said, are very graceful in form—tall, and elegant in their movements, wearing their hair very long and smooth; they also bind their heads with handsome worked handkerchiefs, which, from a distance, look like silk or gauze; others use the same material in a longer form, wound round them so as to cover them like trousers, and this is done by both the men and the women. These people are of a whiter skin than any that I have seen in the Indies. It is the fashion among all classes to wear something at the breast, and on the arms, and many wear pieces of gold hanging low on the bosom. Their canoes are larger, lighter, and of better build than those of the islands which I have hitherto seen, and in the middle of each they have a cabin or room, which I found was occupied by the chiefs and their wives. I called this place "Jardines," that is, "the Gardens," for the place and the people corresponded with that appellation. I made many inquiries as to where they found the gold, in reply to which, all of them directed me to an elevated tract of land at no great distance, on the confines of their own country, lying to the westward; but they all advised me not to go there, for fear of being eaten, and at the time I imagined that by their description they wished to imply that they were cannibals who dwelt there, but I have since thought it possible that they meant merely to express that the country was filled with beasts of prey. I also inquired of them where they obtained the pearls? and in reply to this question likewise, they directed me to the westward, and also to the north, behind the country they occupied. I did not put this information to the test, on account of the provisions, and the weakness of my eyes, and because the large ship that I had with me was not calculated for such an undertaking. The short time that I spent with them was all passed in putting questions; and at evening, as I have already said, we returned to the ships, upon which I weighed anchor and sailed to the westward. I proceeded onward on the following day, until I found that we were only in three fathoms water; at this time I was still under the idea that it was but an island, and that I should be able to make my exit by the north. Upon which I sent a light caravel in advance of us, to see whether there was any exit, or whether the passage was closed; the caravel proceeded a great distance, until it reached a very large gulf, in which there appeared to be four smaller gulfs, from one of which debouched a large river; they invariably found ground at five fathoms, and a great quantity of very fresh water; indeed, I never tasted any equal to it. I felt great anxiety when I found that I could make no exit, either by the north, south, or west, but that I was enclosed on all three sides by land; I therefore weighed anchor, and sailed in a backward direction, with the hope of finding a passage to the north by the strait, which I have already described; but I could not return along the inhabited part where I had already been, on account of the currents, which drove me entirely out of my course. But constantly, at every headland, I found the water sweet and clear, and we were carried eastward very powerfully toward the two straits already mentioned; I then conjectured that the currents and the overwhelming mountains of water which rushed into these straits with such an awful roaring, arose from the contest between the fresh water and the sea. The fresh water struggled with the salt to oppose its entrance, and the salt contended against the fresh in its efforts to gain a passage outward. And I formed the conjecture, that at one time there was a continuous neck of land from the island of Trinidad and with the land of Gracia, where the two straits now are, as your Highnesses will see, by the drawing which accompanies this letter. I passed out by this northern strait, and found the fresh water came even there; and when, by the aid of the wind, I was enabled to proceed, I remarked, while on one of the watery billows which I have described, that in the channel the water on the inner side of the current was fresh, and on the outside salt.

When I sailed from Spain to the Indies I found that, as soon as I had passed a hundred leagues westward of the Azores, there was a very great change in the sky and the stars, in the temperature of the air, and in the water of the sea; and I have been very diligent in observing these things. I remarked that from north to south in traversing these hundred leagues from the said islands, the needle of the compass, which hitherto had turned toward the northeast, turned a full quarter of the wind to the northwest, and this took place from the time when we reached that line. At the same time an appearance was presented as if the seashore had been transplanted thither, for we found the sea covered all over with a sort of weed resembling pine branches, and with fruits like that of the mastic tree, so thick that on my first voyage I thought it was a reef, and that the ships could not avoid running aground; but until I reached this line I did not meet with a single bough. I also observed that at this point the sea was very smooth, and that, though the wind was rough, the ships never rolled. I likewise found that within the same line toward the west the temperature was always mild, and that it did not vary, summer or winter. While I was there I observed that the north star described a circle five degrees in diameter; that when its satellites are on the right side, then the star was at its lowest point, and from this point it continues until it reaches the left side, where it is also at five degrees, and then again it sinks until it at length returns to the right side. In this voyage I proceeded immediately from Spain to the island of Madeira, thence to the Canaries, and then to the Cape Verd Isles, and from the Cape Verd Isles I sailed southward, even below the equinoctial line, as I have already described. When I reached the parallel of Sierra Leone, in Guinea, I found the heat so intense and the rays of the sun so fierce that I thought that we should have been burnt; and, although it rained, and the sky was heavy with clouds, I still suffered the same oppression until our Lord was pleased to grant me a favorable wind, giving me an opportunity of sailing to the west, so that I reached a latitude where I experienced, as I have already said, a change in the temperature. Immediately upon my reaching this line the temperature of the sky became very mild, and the more I advanced the more this mildness increased; but I did not find the positions of the stars correspond with these effects. I remarked at this place that when night came on the polar star was five degrees high, and then the satellites were overhead; afterward, at midnight, I found the north star elevated ten degrees, and when morning was advancing, the satellites were fifteen feet below. I found the smoothness of the sea continue, but not so the weeds; as to the polar star, I watched it with great wonder, and devoted many nights to a careful examination of it with the quadrant, and I always found that the lead and line fell to the same point. I look upon this as something new, and I think my opinion will be supported by that of others, and it is a short distance for so great a change to take place in the temperature. I have always read that the world, comprising the land and the water, was spherical, as is testified by the investigations of Ptolemy and others, who have proved it by the eclipses of the moon, and other observations made from east to west, as well as by the elevation of the pole from north to south. But I have now seen so much irregularity, as I have already described, that I have come to another conclusion respecting the earth, namely: that it is not round, as they describe, but of the form of a pear, which is very round except where the stalk grows, at which part it is most prominent; or like a round ball, upon one part of which is a prominence like a woman's nipple, this protrusion being the highest and nearest the sky, situated under the equinoctial line, and at the eastern extremity of this sea— I call that the eastern extremity, where the land and the islands end. In confirmation of my opinion I revert to the arguments which I have above detailed respecting the line, which passes from north to south a hundred leagues westward of the Azores; for in sailing thence westward the ships went on rising smoothly toward the sky, and then the weather was felt to be milder, on account of which mildness the needle shifted one point of the compass; the further we went the more the needle moved to the northwest, this elevation producing the variation of the circle which the north star describes with its satellites; and the nearer I approached the equinoctial line the more they rose, and the greater was the difference between these stars and their circles. Ptolemy and the other philosophers who have written upon the globe thought that it was spherical, believing that this hemisphere was round as well as that in which they themselves dwelt, the centre of which was in the island of Arin, which is under the equinoctial line between the Arabian Gulf and the Gulf of Persia; and the circle passes over Cape St. Vincent, in Portugal, westward and eastward, by Cangara and the

Seras,* in which hemisphere I make no difficulty as to its being a perfect sphere, as they describe; but this western half of the world, I maintain, is like the half of a very round pear, having a raised projection for the stalk, as I have already described, or like a woman's nipple on a round ball. Ptolemy and the others who have written upon the globe had no information respecting this part of the world, which was then unexplored; they only established their arguments with respect to their own hemisphere, which, as I have already said, is half of a perfect sphere. And now that your Highnesses have commissioned me to make this voyage of discovery, the truths which I have stated are evidently proved, because in this voyage, when I was off the island of Hargin,† and its vicinity, which is twenty degrees to the north of the equinoctial line, I found the

---

* Names for Japan and China, according to Major, but he evidently erred, for if this was so, Columbus must have reached the conclusion that he was not in the Indies, while he died firmly convinced that the lands were part of the East Indies.

† Probably Arguin, on the African coast.

people are black, and the land very much burnt; and when after that I went to the Cape Verd Islands, I found the people there much darker still, and the more southward we went the more they approach the extreme of blackness; so that when I reached the parallel of Sierra Leone, where, as night came on, the north star rose five degrees, the people there were excessively black; and, as I sailed westward, the heat became extreme. But after I had passed the meridian or line which I have already described, I found the climate become gradually more temperate, so that when I reached the island of Trinidad, where the north star rose five degrees as night came on, there and in the land of Gracia I found the temperature exceedingly mild; the fields and the foliage likewise were remarkably fresh and green, and as beautiful as the gardens of Valencia in April. The people there are very graceful in form, less dark than those whom I had before seen in the Indies, and wear their hair long and smooth; they are also more shrewd, intelligent and courageous. The sun was then in the sign of Virgo, over our heads and theirs; therefore all this must proceed from the extreme blandness of the temperature, which arises, as I have said, from this country being the most elevated in the world, and the nearest to the sky. On these grounds, therefore, I affirm that the globe is not spherical, but that there is the difference in its form which I have described; the which is to be found in this hemisphere at the point where the Indies meet the ocean, the extremity of the hemisphere being below the equinoctial line. And a great confirmation of this is, that when our Lord made the sun, the first light appeared in the first point of the east, where the most elevated point of the globe is; and, although it was the opinion of Aristotle that the antarctic pole, or the land which is below it, was the highest part of the world, and the nearest to the heavens, other philosophers oppose him, and say that the highest part was below the arctic pole, by which reasoning it appears that they understood that one part of the world ought to be loftier and nearer the sky than the other; but it never struck them that it might be under the equinoctial in the way that I have said, which is not to be wondered at, because they had no certain knowledge respecting this hemisphere, but merely vague suppositions, for no one has ever gone or been sent to investigate the matter until your Highnesses sent me to explore both the sea and the land. I found that between the two straits which, as I have said before, face each other in a line from north to south, is a distance of twenty-six leagues; and there can be no mistake in this calculation, because it was made with the quadrant. I also find that from these two straits on the west, up to the above-mentioned gulf, to which I gave the name of the Gulf of Pearls, there are sixty-eight leagues of four miles to the league, which is the reckoning we are accustomed to make at sea; from this gulf the water runs constantly, with great impetuosity, toward the east, and this is the cause why in these two straits there is so fierce a turmoil from the fresh water encountering the water of the sea. In the southern strait, which I named the Serpent's Mouth,

I found that toward evening the polar star was nearly at five degrees elevation; and in the northern, which I called the Dragon's Mouth, it was at an elevation of nearly seven degrees. The before-mentioned Gulf of Pearls is to the west of the * of Ptolemy, nearly three thousand nine hundred miles, which make nearly seventy equinoctial degrees, reckoning fifty-six and two-thirds to a degree. The Holy Scriptures record that our Lord made the earthly paradise and planted in it the tree of life, and thence springs a fountain from which the four principal rivers in the world take their source, namely: the Ganges, in India, the Tigris and Euphrates in † which rivers divide a chain of mountains, and forming Mesopotamia, flow thence into Persia—and the Nile, which rises in Ethiopia and falls into the sea at Alexandria.

I do not find, nor have ever found, any account by the Romans or Greeks which fixes in a positive manner the site of the terrestrial

---

* Words lacking in original MS.
† Words lacking in original MS.

paradise, neither have I seen it given in any mappe-monde laid down from authentic sources. Some place it in Ethiopia, at the sources of the Nile, but others, traversing all these countries, found neither the temperature nor the altitude of the sun correspond with their ideas respecting it; nor did it appear that the overwhelming waters of the deluge had been there. Some pagans pretended to adduce arguments to establish that it was in the Fortunate Islands, now called the Canaries, etc.

St. Isidore, Bede, Strabo, and the master of scholastic history,* with St. Ambrose, and Scotus, and all the learned theologians agree that the earthly paradise is in the east, etc.

I have already described my ideas concerning this hemisphere and its form, and I have no doubt that if I could pass below the equinoctial line, after reaching the highest point of which I have spoken, I should find a much milder temperature, and a variation in the stars and in the waters; not that I suppose that elevated point to be navigable, nor even that there

* Petrus Comestor, author of the *Historia Scholastica.*

is water there; indeed, I believe it is impossible to ascend thither, because I am convinced that it is the spot of the earthly paradise, whither no one can go but by God's permission; but this land which your Highnesses have now sent me to explore is very extensive, and I think there are many other countries in the south of which the world has never had any knowledge.

I do not suppose that the earthly paradise is in the form of a rugged mountain, as the descriptions of it have made it appear, but that it is on the summit of the spot which I have described as being in the form of the stalk of a pear; the approach of it from a distance must be by a constant and gradual ascent; but I believe that, as I have already said, no one could ever reach the top; I think also that the water I have described may proceed from it, though it be far off, and that, stopping at the place which I have just left, it forms this lake. There are great indications of this being the terrestrial paradise, for its site coincides with the opinion of the holy and wise theologians whom I have mentioned; and moreover, the other evidences agree with the supposition, for I have never either read or heard of fresh water coming in so large a quantity in close conjunction with the water of the sea; the idea is also corroborated by the blandness of the temperature; and if the water of which I speak does not proceed from the earthly paradise, it appears to be still more marvelous, for I do not believe that there is any river in the world so large or so deep.

When I left the Dragon's Mouth, which is the northernmost of the two straits which I have described, and which I so named on the day of our Lady of August, I found that the sea ran so strongly to the westward that between the hour of mass, when I weighed anchor, and the hour of complines I made sixty-five leagues of four miles each; and not only was the wind not violent, but on the contrary very gentle, which confirmed me in the conclusion that in sailing southward there is a continuous ascent, while there is a corresponding descent toward the north.

I hold it for certain that the waters of the

sea move from east to west with the sky, and that in passing this track they hold a more rapid course, and have thus carried away large tracts of land, and that from hence has resulted this great number of islands; indeed, these islands themselves afford an additional proof of it, for all of them, without exception, run lengthwise from west to east, and from the northwest to the southeast, which is in a directly contrary direction to the said winds; furthermore, that these islands should possess the most costly productions is to be accounted for by the mild temperature, which comes to them from heaven, since these are the most elevated parts of the world. It is true that in some parts the waters do not appear to take this course, but this occurs in certain spots, where they are obstructed by land, and hence they appear to take different directions.

Pliny writes that the sea and land together form a sphere, but that the ocean forms the greatest mass, and lies uppermost, while the earth is below and supports the ocean, and that the two afford a mutual support to each other, as the kernel of a nut is confined by its shell. The master of scholastic history, in commenting upon Genesis, says that the waters are not very extensive; and that although, when they were first created they covered the earth, they were yet vaporous like a cloud, and that afterward they became condensed, and occupied but small space, and in this notion Nicolas de Lira agrees. Aristotle says that the world is small, and the water very limited in extent, and that it is easy to pass from Spain to the Indies; and this is confirmed by Avenruyz,* and by the Cardinal Pedro de Aliaco,† who, in supporting this opinion, shows that it agrees with that of Seneca, and says that Aristotle had been enabled to gain information respecting the world by means of Alexander the Great, and Seneca by means of Nero, and Pliny through the Romans; all of them having expended

* Averrhòes, an Arabian philosopher of the twelfth century.
† Better known as Pierre D'Ailly—author of the *Ymago Mundi*, stolen largely from Roger Bacon, and which supplied Columbus with so much of his geographical knowledge.

large sums of money, and employed a vast number of people, in diligent inquiry concerning the secrets of the world, and in spreading abroad the knowledge thus obtained. The said cardinal allows to these writers greater authority than to Ptolemy, and other Greeks and Arabs; and in confirmation of their opinion concerning the small quantity of water on the surface of the globe, and the limited amount of land covered by that water, in comparison of what had been related on the authority of Ptolemy and his disciples, he finds a passage in the third book of Esdras, where that sacred writer says, that of seven parts of the world six are discovered, and the other is covered with water. The authority of the third and fourth books of Esdras is also confirmed by holy persons, such as St. Augustine, and St. Ambrose in his *Exameron*, where he says, " Here my son Jesus shall first come, and here my son Christ shall die!" These holy men say that Esdras was a prophet, as well as Zacharias, the father of St. John, and El Braso Simon; authorities which are also quoted by Francis de Mairones. With respect to the dryness of the land, experience has shown that it is greater than is commonly believed; and this is no wonder, for the further one goes the more one learns. I now return to my subject of the land of Gracia, and of the river and lake found there, which latter might more properly be called a sea; for a lake is but a small expanse of water, which, when it becomes great, deserves the name of a sea, just as we speak of the Sea of Galilee and the Dead Sea; and I think that if the river mentioned does not proceed from the terrestrial paradise, it comes from an immense tract of land situated in the south, of which no knowledge has been hitherto obtained. But the more I reason on the subject, the more satisfied I become that the terrestrial paradise is situated in the spot I have described; and I ground my opinion upon the arguments and authorities already quoted. May it please the Lord to grant your Highnesses a long life, and health and peace to follow out so noble an investigation; in which I think our Lord will receive great service, Spain considerable increase of its greatness, and all Christians much consolation and pleasure, because by this means the name of our Lord will be published abroad.

In all the countries visited by your Highnesses' ships, I have caused a high cross to be fixed upon every headland, and have proclaimed to every nation that I have discovered the lofty estate of your Highnesses, and of your court in Spain. I also tell them all I can respecting our holy faith and of the belief in the holy mother Church, which has its members in all the world; and I speak to them also of the courtesy and nobleness of all Christians, and of the faith they have in the Holy Trinity. May it please the Lord to forgive those who have calumniated and still calumniate this excellent enterprise, and oppose, and have opposed its advancement, without considering how much glory and greatness will accrue from it to your Highnesses throughout all the world. They cannot state anything in disparagement of it, except its expense, and that I have not immediately sent back the ships loaded with gold. They speak this without considering the shortness of the time, and how many difficulties there are to contend with; and that every year there are individuals who singly earn by their deserts out of your Majesties' own household more revenue than would cover the whole of this expense. Nor do they remember that the princes of Spain have never gained possession of any land out of their own country, until now that your Highnesses have become the masters of another world, where our holy faith may become so much increased, and whence such stores of wealth may be derived; for although we have not sent home ships laden with gold, we have, nevertheless, sent satisfactory samples, both of gold and of other valuable commodities, by which it may be judged that in a short time large profits may be derived. Neither do they take into consideration the noble spirit of the princes of Portugal, who so long ago carried into execution the exploration of Guinea, and still follow it up along the coast of Africa, in which one-half of the population of the country has been employed, and yet the King is more determined on the enterprise than ever. The Lord grant all that I have said, and lead them to think deeply upon what I have written; which is not the thousandth part of what might be written of the deeds of princes who have set their minds upon gaining knowledge, and upon obtaining territory and keeping it.

I say all this, not because I doubt the inclination of your Highnesses to pursue the enterprise while you live—for I rely confidently on the answers your Highnesses once gave me by word of mouth—nor because I have seen any change in your Highnesses, but from the fear of what I have heard from those of whom I have been speaking; for I know that water dropping on a stone will at length make a hole. Your Highnesses responded to me with that nobleness of feeling which all the world knows you to possess, and told me to pay no attention to these calumniations; for that your intention was to follow up and support the undertaking, even if nothing were gained by it but stones and sand. Your Highnesses also desired me to be in no way anxious about the expense, for that much greater cost had been incurred on much more trifling matters, and that you considered all the past and future as well laid out; for that your Highnesses believed that our holy faith would be increased, and your royal dignity enhanced, and that they were no friends of the royal estate who spoke ill of the enterprise.

And now, during the despatch of the information respecting these lands which I have recently discovered, and where I believe in my soul that the earthly paradise is situated, the "Adelantado" will proceed with three ships, well stocked with provisions, on a further investigation, and will make all the discoveries he can about these parts. Meanwhile, I shall send your Highnesses this letter, accompanied by a drawing of the country, and your Majesties will determine on what is to be done, and give your orders as to how it is your pleasure that I should proceed: the which, by the aid of the Holy Trinity, shall be carried into execution with all possible diligence, in the faithful service and to the entire satisfaction of your Majesties. *Deo Gratias!*

LETTER TO JUANA DE LA TORRES*

MOST VIRTUOUS LADY: Although it is a novelty for me to complain of the ill-usage of the world, it is, nevertheless, no novelty for the world to practice ill-usage. Innumerable are the contests which I have had with it, and I have resisted all its attacks until now, when I find that neither strength nor prudence is of any avail to me: it has cruelly reduced me to the lowest ebb. Hope in Him who has created us all is my support: His assistance I have always found near at hand. On one occasion, not long since, He supported

* The former nurse of Prince Don John. Major thinks this letter was written when Columbus was nearing Cadiz, which he reached Nov. 25, 1500; but from the question in his mind (shown in the last paragraph), as to whether he was to be tried in the Indies or in Spain, it must clearly have been written in San Domingo, before he started. He was certainly under arrest, and perhaps in chains. The translation is by R. H. Major, and is printed in his *Select Letters of Christopher Columbus*, . . . *London*, 1847. The original text is in Navarrete's *Coleccion de los Viages*, . . . *Madrid*, 1825.

me with His Divine arm, saying: "O man of little faith, arise, it is I, be not afraid." I offered myself with such earnest devotion to the service of the princes, and I have served them with a fidelity hitherto unequaled and unheard of. God made me the messenger of the new heaven and the new earth, of which He spoke in the Apocalypse by St. John, after having spoken of it by the mouth of Isaiah; and He showed me the spot where to find it. All proved incredulous, except the Queen, my mistress, to whom the Lord gave the spirit of intelligence and the necesssary courage, and made her the heiress of all, as a dear and well-beloved daughter. I went to take possession of it in her royal name. All wished to cover the ignorance in which they were sunk, by enumerating the inconveniences and expense of the proposed enterprise. Her Highness held the contrary opinion, and supported it with all her power. Seven years passed away in deliberations, and nine have been spent in accomplishing things truly memorable, and worthy of being preserved in the history of man.

I have now reached that point, that there is no man so vile but thinks it his right to insult me. The day will come when the world will reckon it as a virtue to him who has not given his consent to their abuse. If I had plundered the Indies, even to the country where is the fabled altar of St. Peter's, and had given them all to the Moors, they could not have shown toward me more bitter enmity than they have done in Spain. Who would believe such things in a country where there has always been so much magnanimity? I desire earnestly to clear myself of this affair, if only I had the means of doing so face to face with my Queen. The support which I have found in our Lord and in her Highness has made me persevere; and I would fain cause her to forget a little the griefs which death has occasioned her.* I undertook another voyage to the new heavens and new earth, which had been hidden hitherto; and if these are not appreciated in Spain, like the other parts of the Indies, it is not at all wonderful, since it is to my labors that they

* The death of her son, Prince John.

are indebted for them. The Holy Spirit encompassed St. Peter, and the rest of the twelve, who all had conflicts here below; they wrought many works, they suffered great fatigues, and at last they obtained the victory. I believed that this voyage to Paria would in some degree pacify them, because of the pearls and the discovery of gold in the island of Española. I left orders for the people to fish for pearls, and collect them together, and made an agreement with them that I should return for them; and I was given to understand that the supply would be abundant.

If I have not written respecting this to their Highnesses, it is because I wished first to render an equally favorable account of the gold; but it has happened with this as with many other things; I should not have lost them, and with them my honor, if I had been only occupied about my own private interests, and had suffered Española to be lost, or even if they had respected my privileges and the treaties. I say the same with regard to the gold which I had then collected, and which I have brought in safety, by Divine grace, after so much loss of life, and such excessive fatigues.

In the voyage which I made by way of Paria, I found nearly half the colonists of Española in a state of revolt,* and they have made war upon me until now as if I had been a Moor; while on the other side, I had to contend with no less cruel Indians. Then arrived Hojeda,† and he attempted to put the seal to all these disorders; he said that their Highnesses had sent him, with promises of presents, of immunities, and treaties; he collected a numer-

---

* From the rule of Bartholomew Columbus, who, in the absence of his brother, acted in his stead. The character of the people whom Columbus was called upon to govern can be judged by the requisition for colonists sent by the King and Queen to the "council, auditors, alcaldes, bailiffs, magistrates, knights, esquires, officers, and good men . . . of our kingdoms" ordering them "that all and every person . . . who may have committed, up to the day of the publication of this our letter, any murders and offenses, and other crimes of whatever nature and quality they may be . . . shall go and serve in person in Hispaniola." And yet, because order was not maintained, we are seriously told that Columbus was a bungling and poor governor.

† Alonzo de Hojeda.

ous band, for in the whole island of Española there were few men who were not vagabonds, and there were none who had either wife or children. This Hojeda troubled me much, but he was obliged to retreat, and at his departure he said that he would return with more ships and men, and reported, also, that he had left the Queen at the point of death. In the meanwhile, Vincent Yanez* came with four caravels; and there were some tumults and suspicions, but no further evil. The Indians reported many other caravels to the cannibals, and in Paria; and afterward spread the news of the arrival of six other caravels, commanded by a brother of the alcalde; but this was from pure malice; when at last the hope was lost that their Highnesses would send any more ships to the Indies, and we no longer expected them, and when it was said openly that her Highness (the Queen) was dead. At this time, one Adrian † attempted a new revolt, as he had done before; but our Lord did not permit his

---

* This was the commander of the *Nina* in the first voyage.
† Adrian Mogica.

evil designs to succeed. I had determined not to inflict punishment on any person, but his ingratitude obliged me, however regretfully, to abandon this resolution. I should not have acted otherwise with my own brother, if he had sought to assassinate me, and to rob me of the lordship which my sovereigns had given to my keeping. This Adrian, as is now evident, had sent Don Ferdinand to Xaragua, to assemble some of his partisans, and had some discussions with the alcalde, which ended in violence, but all without any good. The alcalde seized him and a part of his band, and, in fact, executed justice without my having ordered it. While they were in prison they were expecting a caravel, in which they hoped to embark; but the news of what had happened to Hojeda, and which I told them, deprived them of the hope that he would arrive in this ship. It is now six months that I have been ready to leave, to bring to their Highnesses the good news of the gold, and to give up the government of these dissolute people, who fear neither their King nor Queen, but are full of imbecility and malice. I should have been able to pay every one with six hundred thousand maravedis, and for this purpose there were four millions and more of the tithes, without reckoning the third part of the gold.

Before my departure (from Spain), I have often entreated their Highnesses to send to these parts, at my expense, some one charged to administer justice; and since, when I found the alcalde in a state of revolt, I have besought them afresh to send at least one of their servants with letters, because I myself have had so strange a character given to me, that if I were to build churches or hospitals they would call them caves for robbers. Their Highnesses provided for this at last, but in a manner quite unequal to the urgency of the circumstances; however, let that point rest, since such is their good pleasure. I remained two years in Spain without being able to obtain anything for myself, or those who came with me, but this man has gained for himself a full purse: God knows if all will be employed for His service. Already, to begin with, there is a revenue for twenty years, which is, according to man's calculation, an age; and they gather gold in such abundance that there are people who, in four hours, have found the equivalent of five marks; but I will speak on this subject more fully hereafter. If their Highnesses would condescend to silence the popular rumors, which have gained credence among those who know what fatigues I have sustained, it would be a real charity; for calumny has done me more injury than the services which I have rendered to their Highnesses, and the care with which I have preserved their property and their government, have done me good; and, by their so doing, I should be reëstablished in reputation, and spoken of throughout the universe; for the things which I have accomplished are such, that they must gain, day by day, in the estimation of mankind.

In the meanwhile, the commander Bobadilla* arrived at St. Domingo, at which time I was

---

*Francisco de Bobadilla, sent from Spain with a royal commission to endeavor to restore peace and order to the colony.

at La Vega, and the Adelantado at Xaragua, where this Adrian had made his attempt; but by that time everything was quiet, the land was thriving, and the people at peace. The second day of his arrival he declared himself governor, created magistrates, ordered executions, published immunities from the collection of gold and from the paying of tithes; and, in fine, announced a general franchise for twenty years, which is, as I have said, the calculation of an age. He also gave out that he was going to pay every one, although they had not even done the service which was due up to that day; and he further proclaimed, with respect to me, that he would send me back loaded with chains, and my brother also (this he has accomplished); and that neither I, nor any of my family, should return forever to these lands; and, in addition to this, he made innumerable unjust and disgraceful charges against me. All this took place, as I have said, on the very day after his arrival, at which time I was absent at a secure distance, thinking neither of him nor of his coming. Some letters of their Highnesses of which he brought a considerable number signed in blank, he filled up with exaggerated language, and sent round to the alcalde and his myrmidons, accompanying them with compliments and flattery. To me he never sent either a letter or a messenger, nor has he done so to this day. Reflect upon this, madam! What could any man in my situation think? That honor and favor should be granted to him who had given his sanction to plundering their Highnesses of their sovereignty, and who had done so much injury and caused so much mischief?—that he who had defended and preserved their cause through so many dangers, should be dragged through the mire? When I heard this, I thought he must be like Hojeda, or one of the other rebels; but I held my peace, when I learned for certain from the friars that he had been sent by their Highnesses. I wrote to him, to salute him on his arrival, to let him know that I was ready to set out to go to court, and that I had put up to sale all that I possessed. I entreated him not to be in haste on the subject of the immunities; and I assured him that I would shortly yield this, and everything else connected with the government, implicitly into his charge. I wrote the same thing to the ecclesiastics, but I received no answer either from the one or the other. On the contrary, he took a hostile position, and obliged those who went to his residence to acknowledge him for governor, as I have been told, for twenty years. As soon as I knew what he had done with regard to the immunities, I believed it needful to repair so great an error, and I thought he would himself be glad of it; because he had, without any reason or necessity, bestowed upon vagabonds privileges of such importance, that they would have been excessive even for men with wives and children. I published verbally, and by writings, that he could not make use of these grants, because mine had still more power, and I showed the immunities brought by Juan Aguado. All this I did for the purpose of gaining time, that their Highnesses might be informed as to the state of things, and that they might have opportunity to give fresh orders upon everything touching their interests. It is useless to publish such grants in the Indies— all is in favor of the settlers who have taken up their abode there, because the best lands are given up to them; and, at a low estimate, they are worth two hundred thousand maravedis a head for the four years, at which they are taken, without their having given one stroke of the spade or the mattock. I should not say so much if these people were married men; but there are not six among them all whose purpose is not to amass all they can, and then decamp with it. It would be well to send people from Spain, and only to send such as are well known, that the country may be peopled with honest men. I had agreed with these settlers that they should pay the third of the gold and of the tithes; and this they not only assented to, but were very grateful to their Highnesses. I reproached them when I heard they had afterward refused it; they expected, however, to deal with me on the same terms as with the commander, but I would not consent to it. He meanwhile irritated them against me, saying that I wished to deprive them of that which their Highnesses had given them; and strove to make me appear their enemy, in which he succeeded to the full. He induced them to write to their Highnesses, that they should send me no more commissioned as governor (truly, I do not desire it any more for myself, or for any who belong to me, while the people remained unchanged); and to conciliate them, he ordered inquiries to be made respecting me with reference to imputed misdeeds, such as were never invented in hell. But God is above, who, with so much wisdom and power, rescued Daniel and the three children, and who, if he please, can rescue me with a similar manifestation of his power, and to the advancement of his own cause. I should have known well enough how to find a remedy for the evils which I now describe and have been describing as having happened to me since I came to the Indies, if I had had the wish or had thought it decent to busy myself about my personal interest; but now I find myself shipwrecked, because, until now, I have maintained the justice and augmented the territorial dominions of their Highnesses. Now that so much gold is found, these people stop to consider whether they can obtain the greatest quantity of it by theft, or by going to the mines. For one woman they give a hundred castellanos, as for a farm; and this sort of trading is very common, and there are already a great number of merchants who go in search of girls; there are at this moment some nine or ten on sale; they fetch a good price, let their age be what it will. In saying that the commander could not confer immunities, I did what he desired, although I told him that it was in order to gain time until their Highnesses had received information respecting the country, and had given their orders as to the regulations best calculated to advance their interest. I say that the calumnies of injurious men have done me more harm than my services have done me good; which is a bad example for the present as well as for the future. I aver that a great number of men have been to the Indies, who did not deserve baptism in the eyes of God or men, and who are now returning thither. The governor has made every one hostile to me; and it appears from the manner of his acting, and the plans that he has adopted, that he was already my enemy, and very virulent against me when he arrived; and it is said that he has been at great expense to obtain this office; but I know nothing about the matter except what I have heard. I never before heard of any one who was commissioned to make an inquiry, assembling the rebels, and taking as evidence against their governor wretches without faith, and who are unworthy of belief. If their Highnesses would cause a general inquiry to be made throughout the land, I assure you they would be astonished that the island has not been swallowed up. I believe that you will recollect that when I was driven by a tempest into the port of Lisbon (having lost my sails), I was falsely accused of having put in thither with the intention of giving the Indies to the sovereign of that country. Since then, their Highnesses have learned the contrary, and that the report was produced by the malice of certain people. Although I am an ignorant man, I do not imagine that any one supposed me so stupid as not to be aware that, even if the Indies had belonged to me, I could not support myself without the assistance of some prince. Since it is thus, where should I find better support, or more security against expulsion, than in the King and Queen, our Sovereigns? Who, from nothing, have raised me to so great an elevation, and who are the greatest princes of the world, on the land and on the sea. These princes know how I have served them, and they uphold my privileges and rewards; and if any one violates them, their Highnesses augment them by ordering great favor to be shown me, and ordain me many honors, as was shown in the affair of Juan Aguado. Yes, as I have said, their Highnesses have received some services from me, and have taken my son into their household,* which would not have happened with another prince, because where there is no attachment, all other considerations prove of little weight. If I have now spoken

* Diego Columbus.

severely of a malicious slander, it is against my will, for it is a subject I would not willingly recall, even in my dreams. The Governor Bobadilla has maliciously exhibited in open day his character and conduct in this affair; but I will prove without difficulty that his ignorance, his laziness, and his inordinate cupidity, have frustrated all his undertakings. I have already said that I wrote him, as well as to the monks, and I set out almost alone, all our people being with the Adelantado and elsewhere, to remove suspicion; when he heard this, he caused D. Diego to be loaded with irons, and thrown into a caravel; he acted in the same manner toward myself, and toward the Adelantado when he arrived. I have never spoken with him, and to this day he has not permitted any one to hold converse with me, and I make oath that I have no conception for what cause I am made prisoner. His first care was to take the gold that I had, and that without measuring or weighing it, although I was absent; he said he would pay those to whom it was owing, and if I am to believe that which has been reported to me, he reserved to himself the greater part, and sent for strangers to make the bargains. I had put aside certain specimens of this gold, as large as the eggs of a goose or a fowl, and many other sizes, which had been collected in a short space of time, in order to please their Highnesses, and that they might be impressed with the importance of the affair, when they saw a great number of large stones loaded with gold. This gold was the first that, after he had feathered his own nest (which he was in great haste to do), his malice suggested to give away, in order that their Highnesses might have a low opinion of the whole affair; the gold which required melting diminished at the fire, and a chain, weighing nearly twenty marks, disappeared altogether. I have been yet more concerned respecting the affair of the pearls, that I have not brought them to their Highnesses. In everything that could add to my annoyance the governor has always shown himself ready to bestir himself. Thus, as I have said, with six hundred maravedis, I should have paid every one, without occasioning loss to any ; and

I had more than four millions of tithes and constabulary dues, without touching the gold. He made the most absurd gifts, although I believe he began by awarding them to the stronger party; their Highnesses will be able to ascertain the truth on this subject when they demand the account to be rendered them, especially if I may assist at the examination. He is continually saying that there is a considerable sum owing, while it is only what I have already reported, and even less. I have been wounded extremely by the thought that a man should have been sent out to make inquiry into my conduct, who knew that if he sent home a very aggravated account of the result of his investigation, he would remain at the head of the government. Would to God their Highnesses had sent either him or some other person two years ago, for then I know that I should have had no cause to fear either scandal or disgrace; they could not then have taken away my honor, and I could not have been in the position to have lost it. God is just, and He will in due time make known all that has taken place and why it has taken place. I am judged in Spain as a governor who had been sent to a province, or city, under regular government, and where the laws could be executed without fear of endangering the public weal; and in this I received enormous wrong. I ought to be judged as a captain sent from Spain to the Indies, to conquer a nation numerous and warlike, with customs and religion altogether different to ours; a people who dwell in the mountains, without regular habitations for themselves or for us; and where, by the Divine will, I have subdued another world to the dominion of the King and Queen, our Sovereigns; in consequence of which, Spain, that used to be called poor, is now the most wealthy of kingdoms. I ought to be judged as a captain, who, for so many years has borne arms, never quitting them for an instant. I ought to be judged by cavaliers who have themselves won the meed of victory; by gentlemen, indeed, and not by the lawyers; at least as it would have been among the Greeks and Romans, or any modern nation in which exists so much nobility as in

Spain; for under any other judgment I receive great injury, because in the Indies there is neither civil nor judgment seat.

Already the road is opened to the gold and pearls, and it may surely be hoped that precious stones, spices, and a thousand other things will also be found. Would to God that it were as certain that I should suffer no greater wrongs than I have already experienced, as it is that I would, in the name of our Lord, again undertake my first voyage; and that I would undertake to go to Arabia Felix, as far as Mecca, as I have said in the letter that I sent to their Highnesses by Antonio de Torres, in answer to the division of the sea and land between Spain and the Portuguese;* and I would go afterward to the North Pole, as I have said and given in writing to the monastery of the Mejorada.

The tidings of the gold which I said I would give, are, that on Christmas-day, being greatly afflicted and tormented by the wicked Span-

---

* This is a reference to the famous Papal bull, dividing the Indies between Spain and Portugal.

iards and the Indians, at the moment of leaving all to save my life if possible, our Lord comforted me miraculously, saying to me, " Take courage: do not abandon thyself to sadness and fear; I will provide for all; the seven years, the term of the gold, are not yet passed, and in this, as in the rest, I will redress thee." I learned, that same day, that there were twenty-four leagues of land where they found mines at every step, which appear now to form but one. Some of the people collected a hundred and twenty castellanos' worth in one day, others ninety; and there have been those who have gathered the equivalent of nearly two hundred and fifty castellanos. They consider it a good day's work when they collect from fifty to seventy, or even from twenty to fifty, and many continue searching; the mean day's work is from six to twelve, and those who get less are very dissatisfied. It appears that these mines, like all others, do not yield equally every day; the mines are new, and those who collect their produce are inexperienced. According to the judgment of everybody here, it seems that if all Spain were to come over, every individual, however inexpert he might be, would gain the equivalent of at least one or two castellanos in a day; and so it is up to the present time. It is certain that any man who has an Indian to work for him collects as much, but the working of the traffic depends upon the Spaniard. See, now, what discernment was shown by Bobadilla when he gave up everything for nothing, and four millions of tithes without any reason, and even without being asked to do so, and without first giving notice to their Highnesses of his intention; and this is not the only evil which he has caused. I know, assuredly, that the errors which I may have fallen into have been done without the intention to do wrong, and I think that their Highnesses will believe me when I say so; but I know and see that they show mercy toward those who intentionally do injury to their service. I, however, feel very certain that the day will come when they will treat me much better; since, if I have been in error, it has been innocently and under the force of circumstances, as they will shortly understand beyond all doubt. I, who am their creature, and whose services and usefulness they will every day be more willing to acknowledge. They will weigh all in the balance, even as, according to Holy Scripture, it will be with the evil and the good at the day of judgment. If, nevertheless, their Highnesses ordain me another judge, which I hope will not be the case, and if my examination is to be holden in the Indies, I humbly beseech them to send over two conscientious and respectable persons at my expense, who would readily acknowledge that, at this time, five marks of gold may be found in four hours; be it, however, as it may, it is highly necessary that their Highnesses should have this matter inquired into. The governor, on his arrival at Española, took up his abode in my house, and appropriated to himself all that was therein. Well and good; perhaps he was in want of it; but even a pirate does not behave in this manner toward the merchants that he plunders. That which grieved me most was the seizure of my papers, of which I have never been able to recover one;

and those that would have been most useful to me in proving my innocence are precisely those which he has kept most carefully concealed. Behold the just and honest inquisitor! I am told that he does not at all confine himself to the bounds of justice, but that he acts in all things despotically. God our Saviour retains His power and wisdom as of old; and, above all things, He punishes ingratitude.

PRIVILEGES OF COLUMBUS *

The declaration of what belongs, and should, and ought to belong to the Admiral of the Indies, in virtue of the capitulation and agreement, which he entered into with their Highnesses, which is the title and right that the Admiral and his descendants have upon the islands and mainland in the ocean, is as follows:

CHAPTER I

FIRST, by the first article their Highnesses appointed him their Admiral of the islands and mainland discovered and to be discovered in the ocean, with the preëminences, and according to and in the manner that the Admiral of the sea of Castile holds and enjoys his Admiralty in his district.

* Probably prepared in 1501, after Columbus had been freed from arrest, yet not restored to his offices in the "Indies." A copy of it was sent to Nicolo Oderigo before March 21, 1502. It is a further and more elaborate argument on the question discussed in the paper printed at page 75 of this volume. The translation is in the *Memorials of Columbus*, ... *London*, 1823. The original text is in the *Codice Diplomatico Colombo-Americano*, ... *Genova*, 1823.

By the declaration of this it is to be observed, that the Admiral of Castile, in virtue of his privilege, has the third part of whatever is acquired, or he may acquire in the sea; for the same reason, therefore, the Admiral of the Indies ought to have the third part of them, and of whatever is acquired in them.

For inasmuch as the Admiral of Castile enjoys no third except of what is acquired in that sea of which he is Admiral, the Admiral of the Indies ought to have a third of them, and of whatever is acquired by land in them.

The reason of this is, because their Highnesses ordered him to acquire islands and mainland, and designated him especially Admiral of them; and from them and in them he is to receive his reward, being Admiral of, and having acquired them with great peril, contrary to the opinion of everybody.

CHAPTER II

By the second chapter, their Highnesses appointed him their Viceroy and Governor-General of all the said islands and mainland,

PRIVILEGES OF COLUMBUS 179

with the faculty of enjoying all the offices which appertain to the government; excepting that, out of three, one should be appointed by their Highnesses; and afterward their Highnesses conferred upon him a fresh grant of the said offices in the years '92 and '93, by privilege granted, without the said exception.

The declaration of this is, that the said offices of Viceroy and Governor belong to the said Admiral, with the power of appointing all the officers to the offices and magistracies of the said Indies, since their Highnesses, as a reward, and, as it were, as a payment for the labor and pains incurred by the said Admiral in discovering and acquiring possession of the said islands, conferred upon him the grant of the said offices and government with the said power.

For it is very evident, that in the beginning the said Admiral would not have exposed himself, nor would any other person have exposed himself, to so great a risk and danger, if their Highnesses had not granted to him the said offices and government as a reward and recompense for such an undertaking.

Which their Highnesses justly bestowed upon him, in order that the said Admiral might, in preference to every other person, be benefited, honored, and elevated through the same means by which he had rendered them so signal a service. For very little, or scarcely any, honor would accrue to the Admiral, whatever other recompense he might have, if in that land, acquired by him with such difficulty, their Highnesses were to appoint another superior; and as he was appointed to them for such just causes, the said offices and government in justice belong to the said Admiral.

And as the said Admiral was peacefully executing, in the service of their Highnesses, the said offices in the said Indies, he was unjustly deprived of the possession of them,* contrary to all law and reason, without being cited, heard, or convicted; by which the said Admiral declares he received every injury, considerable personal dishonor, and loss of property; and

---

* Referring to the appointment and seizure of the government of Hispaniola by Bobadilla. See *ante.*

this clearly appears by the said Chapter, for the following reasons:

Because the said Admiral could not be deprived nor dispossessed of his foresaid offices, never having committed or done anything against their Highnesses, for which he should legally forfeit his property. Supposing, however, that such cause existed (which God forbid!) the said Admiral ought, first of all, to have been cited and called, heard and convicted, according to law.

And by dispossessing him without just cause, the said Admiral experienced great injury and great injustice; and their Highnesses had no right to inflict it upon him.

For their Highnesses conferred upon him the said offices and government of the foresaid land as a compensation for services and labor in gaining possession of it, whence he acquired a just interest in, and perpetual title to, the foresaid offices; and as he was unjustly dispossessed of them, the said Admiral ought, first of all, to be reëstablished in the said offices, and in his honor and dignity.

And with respect to the damage he has received, which, according to the averment of the Admiral, is of considerable amount, as by his persevering industry he was finding out and discovering in the said Indies a great quantity of gold, pearls, spices, and other articles of great value; let the Admiral himself declare upon oath the amount of the damage, and for this let him be indemnified according to law.

Which indemnification ought to be made by the person which unjustly dispossessed him of all his property; being obliged to it both by divine and human laws, for having exceeded the bounds of the power entrusted to him by their Highnesses.

And such indemnification and restoration into the foresaid offices, property, and honors, ought to be the more promptly performed, in proportion to the injustice in depriving him of the same.

For it is absolutely incredible, nor could any one believe, that their Highnesses could approve that a man so industrious, who came so great distance to render such signal and great services to their Highnesses as he has done by his industry and person, by which he merited still greater fortune, should be, through the malignity of the envious, deprived of every recompense.

Having such reason to believe himself bound by affection to their Highnesses, and so well established in their good graces, the said Admiral and all the world believed that it was impossible for any calumniators to make him lose the reward of so many services; much less to excite anger in the breasts of their Highnesses, to make them ruin him whose services and merits they had acknowledged; at a time when the said Admiral was confident of rendering every day, and did render, great services to their Highnesses, promoting by his industry the present advantage of the said islands, and exercising his powers, along with his officers, for their population and prosperity.

And this no other person would have done, nor will do; inasmuch as the Indians being entirely unprotected, if he had not previously governed, those who now possess the government, being anxious to enrich themselves during their administration, will not look to that in future, as the said Admiral did, who looked to their permanent interest, and, depending upon the honor and profit that would result from the good government and protection of the Indians (who form the principal riches of it), attended not in the least to his present advantage.

CHAPTER III

By the Third Chapter their Highnesses conferred upon him a grant of the tenth part of whatever might be bought, found, or existed within the limits of the foresaid Admiralty, deducting the charges of them.

The meaning of this is, that the foresaid Admiral is to have the tenth of whatever might exist or be found in the said Indies and mainland of the ocean, by any persons whatsoever, singly or jointly, for the advantage of their Highnesses, or of whatever other persons to whom they may have made a grant of it, or of part of it, deducting the expenses which the said persons or their Highnesses may have incurred.

And their Highnesses cannot in justice grant either the whole or any part of the profit of the said Indies to any person whatsoever, in prejudice of the said tenth, without their first having to pay, and paying the full tenth thereof to the said Admiral.

For their Highnesses, by making such grants, destroy or diminish that which they formerly conferred upon the said Admiral, leaving it much diminished and dismembered, without an adequate indemnification.

As the grant conferred upon the said Admiral of the said tenth, was given to him before he discovered the said Indies, and was given and granted as an assistance, reward and recompense which he had deserved for that service.

And even supposing that their Highnesses, in conformity to an agreement or condition, or in any other manner, were to give the half, or any other part of the gains to any persons who might be inclined to take upon them the labor and expenses of such an adventure; even in that case the said Admiral ought still to have the tenth of the profits thereof, and of what has been spent by such persons as well as on the principal part of their Highnesses; since both the one and the other are true and principal gain, and are derived from his Admiralty of the Indies.

CHAPTER IV

ACCORDING to the tenor of the Fourth Chapter, their Highnesses granted to the foresaid Admiral the civil and criminal jurisdiction over every dispute in law connected with the foresaid Indies, and the cognizance of them here, in the parts and places comprehended within the jurisdiction of the Admiral of Castile (it being just).

As an explanation of the judicial power belonging to the Admiral, the latter asserts that the foresaid jurisdiction belongs to him as one of the principal preëminences, and, as it were, the arm of the body of his Admiralty, without which it would be very difficult for him to regulate the said Admiralty, or, properly speaking, it would be altogether useless, because the said jurisdiction is the very essence that honors, animates and sustains the other members of the body of the said Admiralty.

Moreover, that the said cognizance belongs to him, as well in the ports and bays of this kingdom as in the said islands and mainland of which he is the Admiral; for if he enjoyed the foresaid jurisdiction only in the courts there, without including in it the causes that emanate from hence, all the contracting parties being natives of this country, and all the traffic and commerce proceeding from hence, his jurisdiction would be almost null, because the individuals who go over to the said Indies go there only for the purpose of trafficking; but the contracts and agreements of the companies remain here, upon which, on their return, law-suits arise; and the causes of such law-suits proceed from transactions in the traffic and commerce which have been carried on within his Admiralty.

But, even if that Article did not exist, in

which express mention is made of the said jurisdiction, it is clear that from the time their Highnesses established the office of the Admiralty of Castile, conjointly with the said Admiralty, they conferred upon the said Admiral the grant of the said jurisdiction with the foresaid comprehension, as the Admiral of the sea of Castile holds, as the principal preëminence of his Admiralty, the jurisdiction of all civil and criminal law-suits appertaining to it; which jurisdiction comprehends all the ports and bays of this country, although out of his Admiralty.

And as to the question, whether it was just to grant him such powers, the foresaid Admiral asserts that their Highnesses could justly confer upon him as kings and sovereigns, lords who have absolute power over all, and to whom only such appointment belongs.

And their Highnesses, in conferring the foresaid office upon the said Admiral, with the foresaid comprehension, did no injury to any person, nor affected any one's interests, because his said Admiralty and its jurisdiction, and the Indies and countries over which it is established, were lately and miraculously discovered, united, and brought under the dominion of Castile.

Moreover, the law-suits emanating from the said Admiralty, on account of the great distance and separation of the countries over which it is established, and being very far from the spot to which the merchants of this country resort, there would be great inconvenience in dividing and separating them from the lawsuits appertaining to this country; and by dividing and separating the cognizance of them, no jurisdiction whatsoever could take place.

And as their Highnesses, without injury to any individual, by their sovereign power did justly make such provision, it is very certain that by it no injustice is committed; because naturally two contraries cannot govern the same subject; that, on the contrary, so foreign are they, and* from existing in one subject, that by the species of one we arrive at a knowledge of the quality of the other: therefore, it may be concluded that the said provision is just.

* Words lacking in original MS.

Even the person of the Admiral proves the justice of the said provision; for, taking into consideration the quality of the said West Indies, unknown to all the world, it was necessary to place there a judge of certain experience, in order to execute just judgments; who was there, then, who possessed greater experience, or more profound knowledge of the nature of the law-suits connected with them, than that Admiral who has constantly resided in them, and miraculously found them through his great skill and knowledge of the sea, and by exposing himself to the innumerable dangers of the said sea?

CHAPTER V

By the fifth chapter their Highnesses grant to the said Admiral the power of contributing the eighth part of any equipment.

The true meaning of this is, that the said Admiral is to have the eighth of whatever articles, in whatsoever manner they may be transported to the foresaid Indies, although it were for the profit of their Highnesses, or of any other person whatsoever, deducting the eighth of the expenses, *pro rata.*

For it must be known that the Admiral contributed his eighth part, and almost half the expense, of the first fleet, by which the Indies were acquired; by which he obtained a perpetual title to the said eighth, on account of the produce of the said expedition being everlasting.

Moreover: as he originally went expressly to acquire islands and mainland, which are unchangeable things, it cannot be explained in what manner he would derive any advantage of enjoying the eighth, if it were not understood that movable things were the scope of the said equipment, as is clearly apparent.

And although the said Admiral in the first expedition did not bring back any movables from the said Indies, which formed the produce and gain of it, he afterward, however, brought the said islands and mainland under the dominion of their Highnesses, and left them peacefully as their own; and, therefore, it is likewise understood, that he consigned and made over to their Highnesses all the movables which then and at all future times should be found in them; wherefore from that time forward their Highnesses could peaceably send for all such things, as their own, whatever person they judged proper.

Allowing, however, that the said Admiral by his contribution to the first expedition had not acquired a perpetual right to the foresaid eighth, nevertheless, as their Highnesses are under the necessity of fitting out vessels to enjoy the profit of the said Indies, they cannot in justice prevent him from concurring in the said expense, and receiving the eighth of the profits; and as the expeditions must continually go on, because the produce of the Indies is continual, the foresaid eighth must forever belong to him.

And as it may be said, that such eighth belongs to him out of the profits of the merchandise alone, because it is expressed in the article of traffic and commerce, that merchandise is understood, the truth is, that the said eighth of all the movables of the Indies belongs specifically to the foresaid Admiral, because the said words *traffic* and *commerce* comprehend every kind of articles that may in any manner or at any time exist.

For the said word *traffic* is the skill or diligence that is employed in obtaining the object of all commerce; and, finally, the traffic or method that was adopted by the said Admiral toward the possessors of the said Indies which he went to acquire, in order to succeed in his intention, which was to acquire them: and as he acquired them, whatever is obtained from them is exactly what ought to be divided as the true produce of such commerce.

And this other word, *negotiation* (commerce) comes from *negotium*, which means *nega otium*, *quia negotium est quasi nega otium;* so that it is generally understood for every kind of thing whatsoever, and on that very account comprehends every kind of movable things that are to be found in the said Indies.

And even supposing that the foresaid word were not equivocal, and had the precise signification of merchandise, it being true that the said Indies and mainland, and particularly Española, were acquired by the said Admiral rather by gifts of merchandise than by force of arms, the said Indies, with all their products, may be justly said to be *mercadas* (purchase), and hence, *mercaderia*, because from *mercar* is derived the said word *mercaderia*.

Moreover: even though the said Admiral had acquired by force of arms the said Indies, and their Highnesses had sent him expressly for the purpose of trafficking, nevertheless he would not lose his right to the foresaid eighth of them; because the movables that are found in them, such as gold, pearls, spices, and other articles, are purely and simply *merchandise:* as every movable article that can be purchased (excepting consecrated articles) is to be looked upon as merchandise according to the tenor of the laws, which declare: *omnia sunt in commercio nostro.*

Besides: in whatever manner the Admiral might have accomplished the object of the equipment of the fleet, which was the acquisition of the said Indies, the said Admiral had a right to his eighth; for the gains of the sea, and their chances, are exceedingly various, fortunate, uncertain, and unexpected; and whatever results from them must be divided among all, whether it has been obtained by force or by stratagem; such being the usage of all privateers, of which we have innumerable examples.

For if any merchants were jointly to fit out a vessel for the sole purpose of trading in merchandise, and granted the captain permission to contribute a part in the equipment, in order to enjoy a correspondent part in the profit; if, besides trading, he should capture any town, money or vessel of an enemy, it is certain that the same quota of such gain would belong to him as by right he would have in the merchandise; because, although the gain proceeds not from merchandise, it is the actual result obtained in consequence of the equipment of the vessel.

And if by chance a factor of any company trading in any kingdom should obtain the favor of the king of that country by assisting him with loans, or by selling goods to him at a lower price; and if the company should be dissolved, and it should happen that the said king, through friendship, after such dissolution, were to make him a present of anything, the factor would be obliged to divide it entirely with his associates as the real profit obtained through the company, although it has been dissolved long since: and thus it has been decided everywhere, and thus the laws of these kingdoms of their Highnesses do declare.

And the same thing happened not long ago in Portugal to a Florentine, the factor of a considerable company in Florence, who, after having rendered many services to the said King by loans, and furnishing him with other goods, was constrained to give a part to his associates of a present which the King made to him personally, through friendship, although the accounts had been already settled and the company dissolved, it being looked upon as a real profit emanating from the said company.

In like manner a certain Captain Lercar, to whom their Highnesses made a present for the attentions he had shown to the Archduchess, and as a compensation for the carack, which he lost upon the sand-banks, was, by the courts of law in Genoa, condemned to give up a part of it to his associates, as the real profit; he receiving that portion only which belonged to him as captain.

And moreover, if by chance any donation should be made by an intimate friend of the father to one of his sons, although all other presents are regarded as private property, this would nevertheless be assigned to the property held in conjunction with the father; because the father was the cause of it: and many other circumstances happen continually, which might be cited upon this head. But passing them over in silence, it will be sufficient to collect, from all that has been said, that the third of the said Indies and mainland justly belongs to the said Admiral, as well as the eighth and tenth of all movable articles, which in them and within the jurisdiction of this Admiralty at whatever time, by whatsoever person, and in any manner whatsoever may be found; they being the real profit of his foresaid expedition, although he may not have contributed to the others: this having been dwelt upon at length in another writing.

I shall here finish by declaring to their Highnesses, that they conferred the grant of all the said offices upon the Admiral, in the same manner as they are enjoyed by the Admiral of the sea of Castile, and that he should appoint the alguazil and notaries, and order them to execute their duties in his name: and this is conformable to the custom of any knight to whom their Highnesses may have given any commission or office, as may be seen in the case of many in Castile, who take to themselves the income, and cause the duty to be performed by one of their servants, or enter into an agreement with some person for that purpose, allowing him a certain portion of the salary: therefore he again supplicates their Highnesses to give him satisfaction, and permit him to execute the duties of the said offices, and enjoy the emoluments thereof; as it was settled by capitulation and special grant.

LETTER TO FERDINAND AND ISABELLA*

MOST serene, and very high and mighty Princes, the King and Queen, our Sovereigns: My passage from Cadiz to the Canaries occupied four days, and thence to the Indies, from which I wrote, sixteen days. My intention was to expedite my voyage as much as possible while I had good vessels, good crews and stores, and because Jamaica was the place to which I was bound. I wrote this in Dominica; and until now my time has been occupied in gaining information.

Up to the period of my reaching these shores I experienced most excellent weather, but the night of my arrival came on with a dreadful

---

* Narrating the events of his fourth voyage to America, on which he sailed May 9, 1502. The letter was written at Jamaica, July 7, 1503. The translation is by R. H. Major, and is printed in his *Select Letters of Columbus*, ... London, 1849. The original text is in Navarrete's *Coleccion de los Viages*, ... *Madrid*, 1825.

tempest, and the same bad weather has continued ever since. On reaching the island of Española I despatched a packet of letters, by which I begged as a favor that a ship should be supplied me at my own cost in lieu of one of those that I had brought with me, which had become unseaworthy, and could no longer carry sail. The letters were taken, and your Highnesses will know if a reply has been given to them. For my part I was forbidden to go on shore; the hearts of my people failed them lest I should take them further, and they said that if any danger were to befall them, they should receive no succor, but, on the contrary, in all probability have some great affront offered them. Moreover every man had it in his power to tell me that the new Governor would have the superintendence of the countries that I might acquire.

The tempest was terrible throughout the night, all the ships were separated, and each one driven to the last extremity, without hope of anything but death; each of them also looked upon the loss of the rest as a matter of certainty. What man was ever born, not even excepting Job, who would not have been ready to die of despair at finding himself as I then was, in anxious fear for my own safety, and that of my son, my brother and my friends, and yet refused permission either to land or to put into harbor on the shores which by God's mercy I had gained for Spain with so much toil and danger?

But to return to the ships: although the tempest had so completely separated them from me as to leave me single, yet the Lord restored them to me in His own good time. The ship which we had the greatest fear for, had put out to sea for safety, and reached the island of Gallega, having lost her boat and a great part of her provisions, which latter loss, indeed, all the ships suffered. The vessel in which I was, though dreadfully buffeted, was saved by our Lord's mercy from any injury whatever; my brother went in the ship that was unsound, and he under God was the cause of its being saved. With this tempest I struggled on till I reached Jamaica, and there the sea became calm, but there was a strong current which carried me as far as the Queen's Garden* without seeing land. Hence as opportunity offered I pushed on for terra firma, in spite of the wind and a fearful contrary current, against which I contended for sixty days, and during that time only made seventy leagues. All this time I was unable to get into harbor, nor was there any cessation of the tempest, which was one continuation of rain, thunder and lightning; indeed it seemed as if it were the end of the world. I at length reached Cape of Gracias a Dios,† and after that the Lord granted me fair wind and tide; this was on the twelfth of September. Eighty-eight days did this fearful tempest continue, during which I was at sea, and saw neither sun nor stars; my ships lay exposed, with sails torn, and anchors, rigging, cables, boats and a great quantity of provisions lost; my people were very weak and humbled in spirit, many of them promising to lead a religious life, and all

---

* A name given to a group of islands south of Cuba.
† In Honduras.

making vows and promising to perform pilgrimages, while some of them would frequently go to their messmates to make confession. Other tempests have been experienced, but never of so long duration or so fearful as this: many whom we look upon as brave men, on several occasions showed considerable trepidation; but the distress of my son* who was with me grieved me to the soul, and the more when I considered his tender age, for he was but thirteen years old, and he enduring so much toil for so long a time. Our Lord, however, gave him strength even to enable him to encourage the rest, and he worked as if he had been eighty years at sea, and all this was a consolation to me. I myself had fallen sick, and was many times at the point of death, but from a little cabin that I had caused to be constructed on deck, I directed our course. My brother was in the ship that was in the worst condition and the most exposed to danger; and my grief on this account was the greater that I brought him with me against his will.

*This was Ferdinand, his natural son.

Such is my fate, that the twenty years of service through which I have passed with so much toil and danger have profited me nothing, and at this very day I do not possess a roof in Spain that I can call my own; if I wish to eat or sleep, I have nowhere to go but to the inn or tavern, and most times lack wherewith to pay the bill. Another anxiety wrung my very heart-strings, which was the thought of my son Diego, whom I had left an orphan, in Spain, and stripped of the honor and property which were due to him, on my account, although I had looked upon it as a certainty, that your Majesties, as just and grateful Princes, would restore it to him in all respects with increase. I reached the land of Cariay, where I stopped to repair my vessels and take in provisions, as well as to afford relaxation to the men, who had become very weak. I myself (who, as I said before, had been several times at the point of death) gained information respecting the gold mines of which I was in search, in the province of Ciamba; and two Indians conducted me to Carambaru, where the people (who go naked) wear golden mirrors round their necks, which they will neither sell, give, nor part with for any consideration. They named to me many places on the seacoast where there were both gold and mines. The last that they mentioned was Veragua, which was five-and-twenty leagues distant from the place where we then were. I started with the intention of visiting all of them, but when I had reached the middle of my journey I learned that there were other mines at so short a distance that they might be reached in two days. I determined on sending to see them. It was on the eve of St. Simon and St. Jude, which was the day fixed for our departure; but that night there arose so violent a storm that we were forced to go wherever it drove us, and the Indian who was to conduct us to the mines was with us all the time. As I had found everything true that had been told me, in the different places which I had visited, I felt satisfied it would be the same with respect to Ciguare, which, according to their account, is nine days' journey across the country westward: they tell me there is a great quantity of gold there, and that the inhabitants wear coral ornaments on their heads, and very large coral bracelets and anklets, with which article also they adorn and inlay their seats, boxes and tables. They also said that the women there wore necklaces hanging down to their shoulders. All the people agree in the report I now repeat, and their account is so favorable that I should be content with the tithe of the advantages that their description holds out. They are all likewise acquainted with the pepper-plant; according to the account of these people, the inhabitants of Ciguare are accustomed to hold fairs and markets for carrying on their commerce, and they showed me also the mode and form in which they transact their various exchanges; others assert that their ships carry guns, and that the men go clothed and use bows and arrows, swords and cuirasses, and that on shore they have horses, which they use in battle, and that they wear rich clothes and have most excellent houses. They also say that the sea surrounds Ciguare, and that at ten days' journey from thence is the river Ganges; these lands appear to hold the same relation to Veragua, as Tortosa to Fontarabia, or Pisa to Venice. When I left Carambaru, and reached the places in its neighborhood, which I have above mentioned as being spoken of by the Indians, I found the customs of the people correspond with the accounts that had been given of them, except as regarded the golden mirrors: any man who had one of them would willingly part with it for three Hawk's-bells, although they were equivalent in weight to ten or fifteen ducats. These people resemble the natives of Española in all their habits. They have various modes of collecting the gold, none of which will bear comparison with the plans adopted by the Christians.

All that I have here stated is from hearsay. This, however, I know, that in the year ninety-four I sailed twenty-four degrees to the westward in nine hours, and there can be no mistake upon the subject, because there was an eclipse; the sun was in Libra, and the moon in

Aries. What I had learned by the mouth of these people I already knew in detail from books. Ptolemy thought that he had satisfactorily corrected Marinus, and yet this latter appears to have come very near the truth.* Ptolemy places Catigara at a distance of twelve lines to the west of his meridian, which he fixes at two degrees and a third above Cape St. Vincent, in Portugal. Marinus comprises the earth and its limits in fifteen lines, and the same author describes the Indus in Ethiopia as being more than four-and-twenty degrees from the equinoctial line, and now that the Portuguese have sailed there, they find it correct.† Ptolemy says also that the most southern land is the first boundary, and that it does not go lower down than fifteen degrees and a third. The world is but small; out of seven divisions

---

* This was in reference to the diameter of the earth, and to the extent of the Indies. It is needless to mention that it was the erroneous theories on this subject which induced Columbus to believe that he could reach the Indies by sailing westward, and led him to make the attempt.

† The expedition under Vasco de Gama, which sailed in 1497, and reached India *via* the Cape of Good Hope.

of it the dry part occupies six, and the seventh is entirely covered by water. Experience has shown it, and I have written it with quotations from the Holy Scripture, in other letters, where I have treated of the situation of the terrestrial paradise, as approved by the holy Church; and I say that the world is not so large as vulgar opinion makes it, and that one degree from the equinoctial line measures fifty-six miles and two-thirds; and this may be proved to a nicety. But I leave this subject, which it is not my intention now to treat upon, but simply to give a narrative of my laborious and painful voyage, although of all my voyages it is the most honorable and advantageous. I have said that on the eve of St. Simon and St. Jude I ran before the wind wherever it took me, without power to resist it; at length I found shelter for ten days from the roughness of the sea and the tempest overhead, and resolved not to attempt to go back to the mines, which I regarded as already in our possession. When I started in pursuance of my voyage it was under a heavy rain, and reaching the harbor of Bastimentos

I put in, though much against my will. The storm and a rapid current kept me in for fourteen days, when I again set sail, but not with favorable weather. After I had made fifteen leagues with great exertions, the wind and the current drove me back again with great fury, but in again making for the port which I had quitted, I found on the way another port, which I named Retrete, where I put in for shelter with as much risk as regret, the ships being in sad condition, and my crews and myself exceedingly fatigued. I remained there fifteen days, kept in by stress of weather, and when I fancied my troubles were at an end, I found them only begun. It was then that I changed my resolution with respect to proceeding to the mines, and proposed doing something in the interim, until the weather should prove more favorable for my voyage. I had already made four leagues when the storm recommenced, and wearied me to such a degree that I absolutely knew not what to do; my wound reopened, and for nine days my life was despaired of; never was the sea so high, so terrific, and so covered with foam; not only did the wind oppose our proceeding onward, but it also rendered it highly dangerous to run in for any headland, and kept me in that sea which seemed to me as a sea of blood, seething like a cauldron on a mighty fire. Never did the sky look more fearful; during one day and one night it burned like a furnace, and every instant I looked to see if my masts and my sails were not destroyed; for the lightnings flashed with such alarming fury that we all thought the ships must have been consumed. All this time the waters from heaven never ceased descending, not to say that it rained, for it was like a repetition of the deluge. The men were at this time so crushed in spirit that they longed for death as a deliverance from so many martyrdoms. Twice already had the ships suffered loss in boats, anchors, and rigging, and were now lying bare without sails.

When it pleased our Lord, I returned to Puerto Gordo, where I recruited my condition as well as I could. I then once more attempted the voyage toward Veragua, although I was by no means in a fit state to undertake it. The wind and currents were still contrary. I arrived at nearly the same spot as before, and there again the wind and currents still opposed my progress; and once again I was compelled to put into port, not daring to encounter the opposition of Saturn with such a boisterous sea, and on so formidable a coast; for it almost always brings on a tempest or severe weather. This was on Christmas-day, about the hour of mass. Thus, after all these fatigues, I had once more to return to the spot from whence I started; and when the new year had set in, I returned again to my task; but although I had fine weather for my voyage, the ships were no longer in a sailing condition, and my people were either dying or very sick. On the day of the Epiphany, I reached Veragua in a state of exhaustion; there, by our Lord's goodness, I found a river and a safe harbor, although at the entrance there were only ten spans of water. I succeeded in making an entry, but with great difficulty; and on the following day the storm recommenced, and had I been still on the outside at that time I should have been unable to enter on account of the reef. It rained without ceasing until the fourteenth of February, so that I could find no opportunity of penetrating into the interior, nor of recruiting my condition in any respect whatever; and on the twenty-fourth of January, when I considered myself in perfect safety, the river suddenly rose with great violence to a considerable height, breaking my cables and the supports to which they were fastened, and nearly carrying away my ships altogether, which certainly appeared to me to be in greater danger than ever. Our Lord, however, brought a remedy as He has always done. I do not know if any one else ever suffered greater trials.

On the sixth of February, while it was still raining, I sent seventy men on shore to go into the interior, and at five leagues' distance they found several mines. The Indians who went with them conducted them to a very lofty mountain, and thence showing the country all round, as far as the eye could reach, told them there was gold in every part, and that, toward the west, the mines extended twenty days' journey; they also recounted the names of the towns and villages where there was more or less of it. I afterward learned that the Cacique Quibian, who had lent these Indians, had ordered them to show the distant mines, and which belonged to an enemy of his; but that in his own territory one man might, if he would, collect in ten days a great abundance of gold. I bring with me some Indians, his servants, who are witnesses of this fact. The boats went up to the spot where the dwellings of these people are situated; and after four hours my brother returned with the guides, all of them bringing back gold which they had collected at that place. The gold must be abundant, and of good quality, for none of these men had ever seen mines before; very many of them had never seen pure gold, and most of them were seamen and lads. Having building materials in abundance, I established a settlement, and made many presents to Quibian, which is the name they gave to the lord of the country. I plainly saw that harmony would not last long, for the natives are of a very rough disposition, and the Spaniards very encroaching; and, moreover, I had taken possession of land belonging to Quibian. When he saw what we did, and found the traffic increasing, he resolved upon burning the houses, and putting us all to death; but his project did not succeed, for we took him prisoner, together with his wives, his children, and his servants. His captivity, it is true, lasted but a short time, for he eluded the custody of a trustworthy man, into whose charge he had been given, with a guard of men; and his sons escaped from a ship, in which they had been placed under the special charge of the master.

In the month of January the mouth of the river was entirely closed up, and in April the vessels were so eaten with the teredo,* that they could scarcely be kept above water. At this time the river forced a channel for itself, by which I managed, with great difficulty, to extricate three of them after I had unloaded them. The boats were then sent back into the

* The mollusk that bores through the bottoms of vessels.

river for water and salt, but the sea became so high and furious, that it afforded them no chance of exit; upon which the Indians collected themselves together in great numbers, and made an attack upon the boats, and at length massacred the men. My brother, and all the rest of our people, were in a ship which remained inside; I was alone, outside, upon that dangerous coast, suffering from a severe fever and worn with fatigue. All hope of escape was gone. I toiled up to the highest part of the ship, and, with a quivering voice and fast-falling tears, I called upon your Highnesses' war-captains from each point of the compass to come to my succor, but there was no reply. At length, groaning with exhaustion, I fell asleep, and heard a compassionate voice address me thus: "O fool, and slow to believe and to serve thy God, the God of all; what did He do more for Moses, or for David his servant, than He has done for thee? From thine infancy He has kept thee under His constant and watchful care. When He saw thee arrived at an age which suited His designs respecting thee, He brought wonderful renown to thy name throughout all the land. He gave thee for thine own the Indies, which form so rich a portion of the world, and thou hast divided them as it pleased thee, for He gave thee power to do so. He gave thee the keys of those barriers of the ocean sea which were closed with such mighty chains; and thou wast obeyed through many lands, and gained an honorable fame throughout Christendom. What more did the Most High do for the people of Israel, when He brought them out of Egypt? or for David, whom from a shepherd He made to be a king in Judea? Turn to Him, and acknowledge thine error—His mercy is infinite. Thine old age shall not prevent thee from accomplishing any great undertaking. He holds under His sway the greatest possessions. Abraham had exceeded a hundred years of age when he begat Isaac; nor was Sarah young. Thou criest out for uncertain help; answer, who has afflicted thee so much and so often, God or the world? The privileges promised by God He never fails in bestowing; nor does

He ever declare, after a service has been rendered Him, that such was not agreeable with His intention, or that He had regarded the matter in another light; nor does He inflict suffering, in order to give effect to the manifestation of His power. His acts answer to His words; and it is His custom to perform all His promises with interest. Thus I have told you what the Creator has done for thee, and what He does for all men. Even now He partially shows thee the reward of so many toils and dangers incurred by thee in the service of others."

I heard all this, as it were, in a trance; but I had no answer to give in definite words, and could but weep for my errors. He who spoke to me, whoever it was, concluded by saying, " Fear not, trust; all these tribulations are recorded on marble, and not without cause." I rose as soon as I could; and at the end of nine days there came fine weather, but not sufficiently so as to allow of drawing the vessels out of the river. I collected the men who were on land, and, in fact, all of them that I could, because there were not enough to admit of one party remaining on shore while another stayed on board to work the vessel. I myself should have remained with my men to defend the buildings I had constructed, had your Highnesses been cognizant of all the facts; but the doubt whether any ships would ever reach the spot where we were, as well as the thought, that while I was asking for succor I might bring succor to myself, made me decide upon leaving. I departed, in the name of the Holy Trinity, on Easter night, with the ships rotten, worn out, and eaten into holes. One of them I left at Belen, with a supply of necessaries; I did the same at Belpuerto. I then had only two left, and they in the same state as the others. I was without boats or provisions, and in this condition I had to cross seven thousand miles of sea; or, as an alternative, to die on the passage with my son, my brother, and so many of my people. Let those who are accustomed to slander and aspersion, ask, while they sit in security at home, " Why didst thou not do so and so under such circumstances?" I wish that they were now embarked in this voyage. I verily believe that another journey of another kind awaits them, if there is any reliance to be placed upon our holy faith.

On the thirteenth of May I reached the province of Mago, which is contiguous to that of Cathay, and thence I started for the island of Española. I sailed two days with a good wind, after which it became contrary. The route that I followed called forth all my care to avoid the numerous islands, that I might not be stranded on the shoals that lie in their neighborhood. The sea was very tempestuous, and I was driven backwards under bare poles. I anchored at an island, where I lost, at one stroke, three anchors; and at midnight, when the weather was such that the world appeared to be coming to an end, the cables of the other ship broke, and it came down upon my vessel with such force that it was a wonder we were not dashed to pieces; the single anchor that remained to me, was, next to the Lord, our only preservation. After six days, when the weather became calm, I resumed my journey, having already lost all my tackle; my ships were pierced with worm-holes, like a bee-hive, and the crew entirely dispersed and downhearted. I reached the island a little beyond the point at which I first arrived at it, and there I stayed to recover myself from the effects of the storm; but I afterward put into a much safer port in the same island. After eight days I put to sea again, and reached Jamaica by the end of June; but always beating against contrary winds, and with the ships in the worst possible condition. With three pumps, and the use of pots and kettles, we could scarcely clear the water that came into the ship; there being no remedy but this for the mischief done by the ship-worm. I steered in such a manner as to come as near as possible to Española, from which we were twenty-eight leagues distant, but I afterward wished I had not done so, for the other ship, which was half under water, was obliged to run in for a port. I determined on keeping the sea in spite of the weather, and my vessel was on the very point of sinking when our Lord miraculously brought us upon land. Who will believe what I now write? I assert that in this letter I have not related one-hundredth part of the wonderful events that occurred in this voyage; those who were with the Admiral can bear witness to it. If your Highnesses would be graciously pleased to send to my help a ship of about sixty-four tons, with two hundred quintals of biscuit and other provisions, there would then be sufficient to carry me and my crew from Española to Spain. I have already said that there are not twenty-eight leagues between Jamaica and Española; and I should not have gone there, even if the ships had been in a fit condition for so doing, because your Highnesses ordered me not to land there. God knows if this command has proved of any service. I send this letter by means of and by the hands of Indians; it will be a miracle if it reaches its destination.

This is the account I have to give of my voyage. The men who accompanied me were a hundred and fifty in number, among whom were many calculated for pilots and good sailors, but none of them can explain whither I went nor whence I came; the reason is very simple: I started from a point above the port of Brazil, and while I was in Española, the storm prevented me from following my intended route, for I was obliged to go wherever the wind drove me; at the same time I fell very sick, and there was no one who had navigated in these parts before. However, after some days, the wind and sea became tranquil, and the storm was succeeded by a calm, but accompanied with rapid currents. I put into harbor at an island called Isla de las Bocas, and then steered for terra firma; but it is impossible to give a correct account of all our movements, because I was carried away by the current so many days without seeing land. I ascertained, however, by the compass and by observation, that I moved parallel with the coast of terra firma. No one could tell under what part of the heavens we were, nor at what period I bent my course for the island of Española. The pilots thought we had come to the island of St. John, whereas it was the land of Mango, four hundred leagues to the westward of where they said. Let them answer and say if they know where Veragua is situated. I assert that they can give no other account than that they went to lands where there was an abundance of gold, and this they can certify surely enough; but they do not know the way to return thither for such a purpose; they would be obliged to go on a voyage of discovery as much as if they had never been there before. There is a mode of reckoning derived from astronomy which is sure and safe, and a sufficient guide to any one who understands it. This resembles a prophetic vision. The Indian vessels do not sail except with the wind abaft, but this is not because they are badly built or clumsy, but because the strong currents in those parts, together with the wind, render it impossible to sail with the bow-line, for in one day they would lose as much way as they might have made in seven; for the same reason I could make no use of caravels, even though they were Portuguese latteens. This is the cause that they do not sail unless with a regular breeze, and they will sometimes stay in harbor waiting for this seven or eight months at a time ; nor is this anything wonderful, for the same very often occurs in Spain. The nation of which Pope Pius writes has now been found, judging at least by the situation and other evidences, excepting the horses with the saddles and poitrels and bridles of gold ; but this is not to be wondered at, for the lands on the sea-coast are only inhabited by fishermen, and moreover I made no stay there, because I was in haste to proceed on my voyage. In Cariay and the neighboring country there are great enchanters of a very fearful character. They would have given the world to prevent my remaining there an hour. When I arrived they sent me immediately two girls very showily dressed ; the eldest could not be more than eleven years of age, and the other seven, and both exhibited so much immodesty that more could not be expected from public women ; they carried concealed about them a magic powder ; when they came I gave them some articles to dress themselves out with, and directly sent them back to the shore. I saw here, built on a mountain, a sepulchre as large as a house, and elaborately sculptured; the body lay uncovered and with the face downward; they also spoke to me of other very excellent works of art. There are many species of animals, both small and large, and very different from those of our country. I had at the time two pigs and an Irish dog, who was always in great dread of them. An archer had wounded an animal like an ape, except that it was larger, and had a face like a man's; the arrow had pierced it from the neck to the tail, which made it so fierce that they were obliged to disable it by cutting off one of its arms and a leg; one of the pigs grew wild on seeing this, and fled; upon which I ordered the *begare* (as the inhabitants called him), to be thrown to the pig, and though the animal was nearly dead, and the arrow had passed quite through his body, yet he threw his tail round the snout of the pig, and then, holding him firmly, seized him by the nape of the neck with his remaining hand, as if he were engaged with an enemy. This action was so novel and so extraordinary that I have thought it worth while to describe it here. There is a great variety of animals here, but they all die of the barra. I saw some very large fowls (the feathers of which resemble wool), lions, stags, fallow-deer and birds.

When we were so harassed with our troubles at sea, some of our men imagined that we were under the influence of sorcery, and even to this day entertain the same notion. Some of the people whom I discovered were cannibals, as was evidenced by the brutality of their countenances. They say that there are great mines of copper in the country, of which they make hatchets and other elaborate articles, both cast and soldered; they also make of it forges, with all the apparatus of the goldsmith, and crucibles. The inhabitants go clothed; and in that province I saw some large sheets of cotton, very elaborately and cleverly worked, and others very delicately penciled in colors. They tell me that more inland, toward Cathay, they have them interwoven with gold. For want of an interpreter we were able to learn but very little respecting these countries, or what they contain. Although the country is very thickly peopled, yet each nation has a very different language; indeed, so much so that they can no more understand each other than we understand the Arabs. I think, however, that this applies to the barbarians on the seacoast, and not to the people who live more inland. When I discovered the Indies I said that they composed the richest lordship in the world: I spoke of gold and pearls and precious stones, of spices and the traffic that might be carried on in them; and because these things were not forthcoming at once, I was abused. This punishment causes me to refrain from relating anything but what the natives tell me. One thing I can venture upon stating, because there are so many witnesses of it, viz., that in this land of Veragua I saw more signs of gold in the first two days than I saw in Española during four years, and that there is not a more fertile or better cultivated country in all the world, nor one whose inhabitants are more timid; added to which, there is a good harbor, a beautiful river, and the whole place is capable of being easily put into a state of defence. All this tends to the security of the Christians and the permanency of their sovereignty, while it affords the hope of great increase and honor to the Christian religion; moreover, the road hither will be as short as that to Española, because there is a certainty of a fair wind for the passage. Your Highnesses are as much lords of this country as of Xerez or Toledo, and your ships that may come here will do so with the same freedom as if they were going to your own royal palace. From hence they will obtain gold, and whereas, if they should wish to become masters of the products of other lands, they will have to take them by force or retire empty-handed, in this country they will simply have to trust their persons in the hands of a savage.

I have already explained my reason for refraining to treat of other subjects respecting which I might speak. I do not state as certain, nor do I confirm even the sixth part of all that I have said or written, nor do I pretend to be at the fountain-head of the information. The Genoese, Venetians, and all other nations that possess pearls, precious stones, and other articles of value, take them to the ends of the world to exchange them for gold. Gold is the most precious of all commodities; gold constitutes treasure, and he who possesses it has all he needs in this world, as also the means of rescuing souls from purgatory, and restoring them to the enjoyment of paradise. They say that when one of the lords of the country of Veragua dies, they bury all the gold he possessed with his body. There were brought to Solomon at one journey six hundred and sixty-six quintals of gold, besides what the merchants and sailors brought, and that which was paid in Arabia. Of this gold he made 200 lances and 300 shields, and the entablature which was above them was also of gold, and ornamented with precious stones: many other things he made likewise of gold, and a great number of vessels of great size, which he enriched with precious stones. This is related by Josephus in his Chronicle de "Antiquitatibus;" mention is also made of it in the Chronicles and in the Book of Kings. Josephus thinks that this gold was found in the Aurea; if it were so, I contend that these mines of the Aurea are identical with those of Veragua, which, as I have said before, extends westward twenty days' journey, at an equal distance from the Pole and the Line. Solomon bought all of it—gold, precious stones and silver—but your Majesties need only to send to seek them to have them at your pleasure. David, in his will, left three thousand quintals of Indian gold to Solomon, to assist in building the Temple; and, according to Josephus, it came from these lands. Jerusalem and Mount Zion are to be rebuilt by the hands of Christians, as God has declared by the mouth of His prophet in the Fourteenth Psalm. The Abbé Joaquim said that he who should do this was to come from Spain; Saint Jerome showed the holy woman the way to accomplish it; and the Emperor of China has, some time since, sent for wise men to instruct him in the faith of Christ. Who will offer himself for this work? Should any one do so, I pledge myself, in the name of God, to convey him safely thither, provided the Lord permits me to return to Spain. The people who have sailed with me have passed through incredible toil and danger, and I beseech your Highnesses, since they are poor, to pay them promptly, and to be gracious to each of them according to their respective merits; for I can safely assert, that to my belief they are the bearers of the best news that ever were carried to Spain. With respect to the gold which belongs to Quibian, the cacique of Veragua, and other chiefs in the neighboring country, although it appears by the accounts we have received of it to be very abundant, I do not think it would be well or desirable, on the part of your Highnesses, to take possession of it in the way of plunder; by fair dealing, scandal and disrepute will be avoided, and all the gold will thus reach your Highnesses' treasury without the loss of a grain. With one month of fair weather I shall complete my voyage. As I was deficient in ships, I did not persist in delaying my course; but in everything that concerns your Highnesses' service, I trust in Him who made me, and I hope also that my health will be reëstablished. I think your Highnesses will remember that I had intended to build some ships in a new manner, but the shortness of the time did not permit it. I had certainly foreseen how things would be. I think more of this opening for commerce, and of the lordship over such extensive mines, than of all that has been done in the Indies. This is not a child to be left to the care of a stepmother.

I never think of Española, and Paria, and other countries, without shedding tears. I thought that what had occurred there would have been an example for others; on the contrary, these settlements are now in a languid state, although not dead, and the malady is incurable, or at least very extensive: let him who brought the evil come now and cure it, if he knows the remedy, or how to apply it; but when a disturbance is on foot, every one is ready to take the lead. It used to be the custom to give thanks and promotion to him who placed his person in jeopardy; but there is no justice in allowing the man who opposed this undertaking to enjoy the fruits of it with his children. Those who left the Indies, avoiding the toils consequent upon the enterprise, and speaking evil of it and me, have since returned with official appointments: such is the case now in Veragua: it is an evil example, and profitless both as regards the business in which we are embarked and as respects the general maintenance of justice. The fear of this, with other sufficient considerations which I clearly foresaw, caused me to beg your Highnesses, previously to my coming to discover these islands and terra firma, to grant me permission to govern in your royal name. Your Highnesses granted my request; and it was a privilege and treaty granted under the royal seal and oath, by which I was nominated Viceroy, and Admiral, and Governor-General of all: and your Highnesses limited the extent of my government to a hundred leagues beyond the Azores and Cape Verd Islands, by a line passing from one pole to the other, and gave me ample power over all that I might discover beyond this line; all which is more fully described in the official document.

But the most important affair of all, and that which cries most loudly for redress, remains inexplicable to this moment. For seven years was I at your royal court, where every one to whom the enterprise was mentioned treated it as ridiculous; but now there is not a man, down to the very tailors, who does not beg to be allowed to become a discoverer. There is reason to believe that they make the voyage only for plunder, and that they are permitted to do so, to the great disparagement of my honor, and the detriment of the undertaking itself. It is right to give God his due, and to receive that which belongs to one's self. This is a just sentiment and proceeds from just feelings. The lands in this part of the world which are now under your Highnesses' sway, are richer and more extensive than those of any other Christian power, and yet, after that I had, by the Divine will, placed them under your high and royal sovereignty and was on the point of bringing your Majesties into the receipt of a very great and unexpected revenue; and while I was waiting for ships to convey me in safety, and with a heart full of joy, to your royal presence, victoriously to announce the news of the gold that I had discovered, I was arrested and thrown, with my two brothers, loaded with irons, into a ship, stripped, and very ill-treated, without being allowed any appeal to justice. Who could believe that a poor foreigner would have risen against your Highnesses, in such a place, without any motive or argument on his side; without even the assistance of any other prince upon which to rely; but on the contrary, amongst your own vassals and natural subjects, and with my sons staying at your royal court? I was twenty-eight years old when I came into your Highnesses' service, and now I have not a hair upon me that is not grey; my body is infirm, and all that was left to me, as well as to my brothers, has been taken away and sold, even to the frock that I wore, to my great dishonor. I cannot but believe that this was done without your royal permission. The restitution of my honor, the reparation of my losses, and the punishment of those who have inflicted them, will redound to the honor of your royal character; a similar punishment also is due to those who plundered me of my pearls, and who have brought a disparagement upon the privileges of my Admiralty. Great and unexampled will be the glory and fame of your Highnesses, if you do this; and the memory of your Highnesses, as just and grateful sovereigns, will survive as a bright example to Spain in future ages. The honest devotedness I have always shown to your Majesties' service, and the so unmerited outrage with which it has been repaid, will not allow my soul to keep silence, however much I may wish it: I implore your Highnesses to forgive my complaints. I am indeed in as ruined a condition as I have related; hitherto I have wept over others; may Heaven now have mercy upon me, and may the earth weep for me. With regard to temporal things, I have not even a blanca for an offering; and in spiritual things, I have ceased here in the Indies from observing the prescribed forms of religion. Solitary in my trouble, sick, and in daily expectation of death, surrounded by millions of hostile savages full of cruelty, and thus separated from the blessed sacraments of our holy Church, how will my soul be forgotten if it be separated from the body in this foreign land? Weep for me, whoever has charity, truth, and justice! I did not come out on this voyage to gain to myself honor or wealth; this is a certain fact, for at that time all hope of such a thing was dead. I do not lie when I say that I went to your Highnesses with honest purpose of heart and sincere 'zeal in your cause. I humbly beseech your Highnesses, that if it please God to rescue me from this place, you will graciously sanction my pilgrimage to Rome and other holy places. May the Holy Trinity protect your Highnesses' lives, and add to the prosperity of your exalted position.

Done in the Indies, in the island of Jamaica,

on the seventh of July, in the year one thousand five hundred and three.

WILL OF COLUMBUS *

In the noble city of Valladolid, on the nineteenth day of the month of May, in the year of the birth of our Saviour Jesus Christ one thousand five hundred and six, before me, Pedro de Hinojedo, clerk of the council of their Highnesses, provincial clerk in their court and chancellery, clerk and notary public in all their kingdoms and seigniories; and of subscribing witnesses: Señor Don Cristobal Colon, Admiral, Viceroy, and Governor-General of the islands and mainland of the Indies discovered and by him designated; being infirm in body, he has declared that whereas he had made his will before a public clerk, that he now did revise and revises the said will, and he did approve and has approved it well, and if necessary he did authorize and has authorized it anew. And now, having enlarged his said will, he had written by his own hand a manuscript which he showed and presented before me the said clerk, which he said was written by his own hand, and signed with his name, that he did authorize and has authorized all that is contained in the said manuscript, before me the said clerk, according to and in the manner and form that is contained in said manuscript, and that all the bequests therein contained shall be executed, and be binding as his latest and final wish. And to execute his said will

* Written Aug. 25, 1505, and executed at Valladolid, May 19, 1506, the day before Columbus died. The original text is in Navarrete's *Coleccion de los Viages*, . . . *Madrid*, 1825.

which he had and has made and authorized, and all that is therein contained, each and every part of it, he did name and has named for his executors and fulfillers of his intention Señor Don Diego Colon, his son, and Don Bartholomew Colon, his brother, and Juan de Porras, treasurer of Vizcaya, that they all three shall execute his will, and all therein contained and in the said manuscript, and all the bequests, legacies and dispositions therein contained. For which purpose he said that he did give, and has given, all the authority requisite, and that he did authorize and has authorized before me the said clerk all that is contained in the said manuscript; and to those present he said that he did request and has requested that they should be witnesses of it. The witnesses who were present, summoned and requested to observe all that is said below, the Bachelor Andres Mirueña and Gaspar de la Misericordia, inhabitants of this said city of Valladolid, and Bartolomé de Fresco, and Alvaro Perez, and Juan Despinosa, and Andrea and Hernando de Vargas, and Francisco Manuel and Fernan Martinez, servants of the said Señor Admiral. The tenor of which said manuscript, as it was written with the own hand of the said Admiral, and signed with his name, *de verbo ad verbum*, is as follows:

WHEN I departed from Spain in the year fifteen hundred and two, I had prepared an ordinance and *mayorazgo** of my property, and in a manner which then seemed

* This must have superseded that printed at page 83, *ante*. It is now unknown.

to me to conform to my wish and to the service of the eternal God, and to my honor and that of my successors: which manuscript I left in the monastery of Cuevas in Seville, in the care of Frey Don Gaspar,* with my other manuscripts and my privileges, and the letters which I possess of the King and of the Queen, our Sovereigns. The which ordinance I approve and confirm by this, which I write for the better accomplishment and declaration of my intention. The which I direct that it be executed in the manner herein specified and contained, that which is provided for by this, is not to be executed by the other, for there is to be no repetition.

I appoint my dear son Don Diego to be my heir of all my property and offices which I hold by right and inheritance, as determined in the mayorazgo, and if he should have no legal male heir, that my son Don Ferdinand shall inherit in the same manner, and if he should have no legal male heir that Don Bartholomew

---

* Gaspar Gorricio, a close friend of Columbus.

my brother shall inherit in the same manner, and likewise if he should have no male heir, that my other brother shall inherit; thus it is intended, from one to the other next of kin of my family, and this continually. And there shall be no female heir unless the males become extinct, and if that should happen let it be the female nearest of kin of my family.

And I direct the said Don Diego, my son, or whoever shall inherit, that they shall neither think nor presume to abridge the said *mayorazgo*, only to increase it and enforce it: it is to be understood that the income which he shall have, with his person and estate, shall be at the service of the King and Queen, our Sovereigns, and for the propagation of the Christian religion.

The King and the Queen, our Sovereigns, when I presented to them the Indies—I say presented, because it is evident that by the will of God, our Sovereign, I gave them, as a thing that was mine, I can say, because I importuned their royal Highnesses for them, which were unknown, and the way hidden from those who spoke concerning them, and for the voyage of discovery excepting to use the information and my person, their royal Highnesses did not expend or desire to expend for the purpose more than a million of maravedis, and it was necessary for me to expend the rest: thus it pleased their royal Highnesses that I should have for my portion, out of the said Indies, islands and mainland which are to the west of a line that they ordered to be drawn between the islands of the Azores, and those of Cape Verd, one hundred leagues, which extends from pole to pole; that I should have for my portion the third and the eighth of all, and also the tenth of whatever is found therein, as is declared more fully by my said privileges and letters of grants.

Because heretofore there has been no revenue received from the said Indies, so that I could separate therefrom the sums which I will mention below, and we hope that by the clemency of our sovereign it may amount to a very large sum; my intention would be and is, that Don Ferdinand, my son, should receive of it one million and a half each year, and Don Bartholomew, my brother, one hundred and fifty thousand maravedis, and Don Diego, my brother, one hundred thousand maravedis, because he belongs to the Church. But this cannot be assured with certainty, because heretofore I have not received nor do I have any known income, as has already been declared.

I say, for the further declaration of the aforesaid, that my wish is that the said Don Diego, my son, shall have the said *mayorazgo* with all my property and offices, in the manner already declared, and as I hold them. *And I say that all the income which he shall receive by reason of the said inheritance, that he shall have ten parts of it every year, and that one part of these ten he shall divide among our relatives who appear to have most need of it*, and poor persons, and in other pious works. And afterward from the remaining nine parts he shall take two and divide them into thirty-five parts, and of these Don Ferdinand,. my son, shall have twenty-seven, and Don Bartholomew shall have five, and Don Diego, my brother, three. And because, as I have already declared, my wish would be that Don Ferdinand, my son, should have one million and a half, and Don Bartholomew one hundred and fifty thousand maravedis, and Don Diego one hundred thousand; and I do not know how it may be assured, because heretofore the said income of the said *mayorazgo* has not been known nor the amount; I say that this order aforesaid should be followed until it shall please our Sovereign that the said two parts of the said nine shall be sufficient and shall amount to such an increase that they shall contain the said million and a half for Don Ferdinand, and one hundred and fifty thousand for Don Bartholomew, and one hundred thousand for Don Diego. And when it shall please God that it may be so, or that if the said two parts, to be understood of the nine aforesaid, shall amount to the sum of one million seven hundred and fifty thousand maravedis, that all the surplus should belong to Don Diego, my son, or whoever shall inherit; and I say and request of the said Don Diego, my son, or of whoever shall inherit, that if the income of this said *mayorazgo* shall grow largely, that it will please me to have the portion aforesaid increased to Don Ferdinand and to my brothers.

I say that this part which I direct to give to Don Ferdinand, my son, that I make of it a *mayorazgo* for him, and that to him shall succeed his eldest son, and in like manner from one to the other perpetually, without the power to sell or exchange or give or abuse in any way, and it shall be in the manner and form which was declared in the other *mayorazgo* which I have made for Don Diego, my son.

I say to Don Diego, my son, and I direct that as soon as he shall have income from the said *mayorazgo* an inheritance sufficient to maintain a chapel, that he shall cause to be appointed three chaplains who shall say three masses every day—one to the honor of the Holy Trinity, another to the Conception of our Lady, and the other for the souls of all the faithful dead, *and for my soul and that of my father and mother and wife.* And that if his wealth is sufficient that he shall enrich the said chapel, and shall increase the supplications and prayers for the honor of the Holy Trinity, and if this can be done in the island Española which God gave to me miraculously, I would be glad to have it there where I invoked it, which is in the plain called of the Conception.

I say and direct to Don Diego, my son, or to whoever shall inherit, that he shall pay all the debts which I leave here in a memorial, in the form therein specified, and all the others which justly seem to be owed by me. And I direct him that he shall have special care for Beatrice Enriquez, the mother of Don Ferdinand, my son, that he shall provide for her so that she may live comfortably, like a person should for whom I have so much regard. And this shall be done for the ease of my conscience, because this has weighed heavily on my soul. The reason therefor it is not proper to mention here. Done on the twenty-fifth of August in the year one thousand five hundred and five.

Christo ferens.

The witnesses who were present and who saw done and authorized all the above said by the said Señor Admiral, according to and in the manner aforesaid: the said Bachelor de Mirueña, Gaspar de la Misericordia, inhabitants of the said city of Valladolid, and Bartolomé de Fresco and Alvar Perez and Juan Despinosa and Andrea and Fernando de Vargas and Francisco Manuel and Fernan Martinez, servants of the said Señor Admiral. And I, the said Pedro de Hinojedo, clerk and notary public aforesaid, together with the said witnesses, to all the aforesaid I was present. And therefore I put here this my notarial mark as such: in testimony of the truth—Pedro de Hinojedo, clerk.

THE END.

INDEX

Aborigines, 42, 48, 53, 54, 57, 60, 64, 69, 84, 117, 119, 125, 155, 173, 204, 213, 222; Description of, 44, 46, 56, 109, 117, 124; Trade with, 40, 57; Religion of, 40, 58; Kings of, 56, 54, 123, 214, 232; Cannibals, 47, 227; Women, 46, 48, 62, 63, 165, 225; Abuse of, 165.
Africa, 111, 148; Cicumnavigation of, 15, 208.
Aguado, Juan, 162, 167.
Alcazar, 111.
Alexander, 110, 111, 144.
America (*see also Indies, West Indies, Western Lands*), Discovery of, 18, 23, 32, 48, 52, 84, 106, 152; Mapping of, 21; Colonies in, 45, 155, 165, 213; Right of Columbus in, 75, 83, 179; Division of, 172; Value of, 50, 64, 109, 147.
———South (*see also Gracia, Trinidad, Veragua*). Discovery, 115, 154; Natives, 117, 123, 136; Exploration, 116, 204; Paradise in, 141.
Anam, province of, 44.
Arabia Felix, 172.
Arabian Gulf, 134.
Arabs, 145.

Arenal, point of, 119, 122.
Arin, island of, 134.
Aristotle, cited, 12, 137, 144.
Aurea, Mines of, 231.
Avenruyz, *see Averrhoes*.
Averrhoes, cited, 144.
Ayte, *see Española*.
Azores, islands, 21, 65, 83, 90, 114, 130, 134.
Bastimentos, harbor of, 209.
Bede, cited, 140.
Belen, 219.
Bobadilla, F. de, 159, 168, 174, 180.
Bocas, Isla de las, 223.
Brazil, 22, 223.
Cadiz, 34, 73, 74, 199.
Canary Islands, 21, 30, 113, 114, 131, 140, 199.
Cangara, 134.
Cannibals, 47, 227.
Carambaru, 204, 207.
Caravels, 30, 45, 215, 212, 225.
Cariay, 204, 225.
Caribee, island of, 115.
Cartography, evidence of, 20.
Castile, kingdom of, 61, 66, 68, 72, 84, 89, 103, 198.
Cathay (China), 16, 35, 53, 220.
Catholic Church, 11, 28, 41, 101, 147.

Catigara, 208.
Cetrefrey, island of, 54.
Ceuta, 111.
Charis, island of, 47.
Chios, 49.
Christian faith, spread of, 28, 41, 49, 103, 229.
Ciamba, province of, 204.
Cibau, province of, 60.
Ciguare, 205, 206, 207.
Cipango (Japan), island of, 84.
Colonies, 45, 155, 165, 113.
Columbus, Bartholomew, 86, 89, 91, 93, 94, 102, 150, 155, 160, 168, 201, 204, 213, 216, 241, 242.
—— Christopher — Varying opinions of, 11.
  Special fitness of, 16.
  Did he first discover America? 18.
  Why his discovery is famous, 23.
  Defects of, 24.
  Letter to Ferdinand and Isabella (1492), 27; (1493) 67; (1498) 105; (1503) 199.
  Gives information to Spanish sovereigns, 28, 106.
  Argues a western route to the Indies, 28, 106.
  His theory treated with contempt, 107, 152.
  Ennobled and privileged, 29, 75, 177, 224.
  Sails from Palos, 28.
  Ships of, 30, 215.
  Letter to Sanchez, 33.
  Discovers West Indies, 34, 52, 84, 108.
  —— Names islands, 35, 53.
  Explores the West Indies, 35, 54.
  Describes natives, 37, 56.
  Settles colony in Española, 45, 61.
  Letter to Santangel, 52.
  Natives believe him heaven-born, 59.
  Encounters great storm, 66, 200.
  Vows a pilgrimage, 66.
  Plans for government of Española, 67.
  Signature of, 72, 90.
  Privileges and rights of, 75, 177, 224.
  Entails his property, 81.
  Complaints against, 108, 112, 151, 158, 164, 166.
  Arms of, 89.
  Encouraged by Spanish sovereigns, 112.
  Sails on third voyage, 112.
  Reaches Cape Verd Islands, 113.
  Discovers South America, 115.
  Adventures in Gulf of Paria, 119, 155.
  Theory of the world's shape, 137.
  Theory of the location of Paradise, 139, 209.
  Letter to Torres, 151.
  Ill-usage of, 151, 158, 168, 180, 236.
  Divine aid to, 83, 152, 173, 216.
  Queen Isabella supports, 152.

INDEX

——Revolts quelled by, 155.
　Troubled by Hojeda, 155, 156.
　Distressing position of, 154.
　Superseded by Bobadilla, 159, 180.
　Placed in chains, 168.
　House of, plundered, 175.
　Fourth voyage of, 199.
　Sickness of, 126, 127, 203, 216, 238.
　Poverty of, 204, 238.
　Adventures on Central American coast, 209.
　Founds settlement at Veragua, 214.
　Reaches Jamaica, 221.
　Will of, 240.
— Diego (son), 86, 89, 91, 93, 96, 98, 100, 102, 104, 167, 191, 204, 241, 242.
— Diego (brother), 86, 89, 93, 94, 246.
— Ferdinand, 86, 93, 94, 98, 157, 203, 242, 248.
Comestor, P., cited, 140.
Corunna, 60.
Cosco, Leander de, 33.
Cuba, *see Juana.*
D'Ailly, P., cited, 144.
Dead Sea, 146.
Deza, Fray D. de, 107.
Dominica, 199.
Dragon's Mouth, 139, 142.
Egypt, 217.
Enchanter, 225.
Enriquez, B., 248.
Esdras, Book of, cited, 145.
Española (*see also Navidad, Xaragua*), 37, 38, 44, 47, 54, 60, 63, 66, 72, 81, 84, 85, 103, 108, 113, 154, 156, 159, 175, 200, 207, 220, 222, 223, 228, 233, 248; First settlement at, 45, 49, 61, 64, 85, 155; Government of, 67; Colonists of, 45, 155, 165; Revolts in, 155, 156; Bobadilla seizes government of, 159, 180.
Ethiopia, 46, 139, 140, 208.
Euphrates, 139.
Ferdinand of Spain (*see also Spain*), 18, 33, 67, 75, 83, 97, 105, 158, 199; Letters of, 30, 161; Letters to, 27, 67, 105, 199.
Fernandina, island of, 35, 53.
Fortunate islands, 140.
Gadibus, 34.
Galea, cape, *see Galeota.*
Galeota, cape, 116.
Galilee, sea of, 146.
Gallega, island of, 201.
Gama, Vasco da, 208.
Ganges, 139, 207.
Genoa, 16, 88, 98, 99, 100, 102, 197, 233.
Gold, 14, 38, 56, 61, 68, 108, 111, 123, 126, 154, 157, 163, 165, 168, 173, 175, 207, 213, 228, 230.
Gomera, 34.
Gracia, 119, 122, 129, 136, 146.
Gracias a Dios, cape, 202.
Granada, 28, 30.
Great Inagua, 35, 53.
Greece, 64.
Greeks, 111, 139, 145, 171.
Guanahani, island of, 35, 53.
Guinea, 62, 111, 131, 148.
Haiti, *see Española.*
Hargin, island of, 135.

254 INDEX

Hispaniola, *see Española.*
Hojeda, A. de, 155, 156, 157, 161.
Honduras, 202.
Indies (*see also West Indies, Cathay, Cipango*), 28, 30, 34, 47, 52, 58, 63, 66, 77, 79, 81, 83, 89, 100, 113, 114, 126, 130, 136, 137, 144, 153, 156, 165, 166, 171, 172, 175, 179, 184, 186, 190, 194, 197, 199, 217, 233; Westerly route to, 15, 18, 29; African route to, 15, 18, 208; European interest in, 14, 28, 232; Division of, 172.
Indians, *see Aborigines.*
Isabella of Spain (*see also Spain, Castile*), 18, 33, 67, 75, 83, 97, 105, 153, 158, 199; Letters of, 30, 31, 161; Letters to, 27, 67, 105, 199.
Isabella, island of, 35, 53.
Jamaica, island of, 85, 199, 201, 222.
Jerusalem, 99, 100, 110; Columbus plans conquest of, 99.
John, Prince Don, of Spain, 28, 88, 151, 153.
Juan, Prince, *see John.*
Juana, island of, 35, 36, 37, 38, 44, 53, 54, 56, 60, 108.
Kooblai Khan, 28, 232.
La Vega, 160.
Lira, Nicolas de, cited, 144.
Lisbon, Columbus at, 51, 66, 166.
Little Inagua, island of, 35, 53.
Madeira, island of, 112, 131.
Mago, province of, 220.

Mahomet, 29.
Mairones, Francis de, cited, 145.
Mango, land of, 224.
Map-makers, testimony of, 21.
Marchena, Fray J. Poe, 107.
Marinus, cited, 2.
Matenin, island of, 48.
Mecca, 172.
Mejorada, 172.
Mesopotamia, 139.
Mozica, Adrian, 156, 157.
Monicongos, 84.
Moors, 27, 111, 153, 155.
Navidad del Senor, settlement of, 45, 49, 61, 64.
Nero, Cæsar, 110, 144.
Nile River, 110, 139, 140.
Nina, caravel, 30, 45.
North Caico, island of, 35, 53, 103–4.
North Pole, 172.
Orinoco, 119.
Palos, 17, 30.
Paria, 119, 154, 155, 156, 233.
Pearls, 123, 127, 154, 228, 230; Gulf of, 138.
Pedro de Aliaco, cardinal, cited, 144.
Perez, Alonzo, 115.
Persia, 139; Gulf of, 134.
Pinta, caravel, 30, 45.
Plato, cited, 12.
Pliny, cited, 12, 143, 144.
Pope of Rome, 28, 33, 172, 225.
Porras, J. de, 241.
Portugal, 15, 66, 111, 134, 148, 172, 196.
Ptolemy, 133, 135, 139, 145, 208.
Puerto Gordo, 211.

INDEX 255

Quibian, Cacique, 214, 215, 233.
Retrete, 210.
Romans, 111, 139, 144, 171.
Sanchez, Raphael, Letter to, 33.
St. Ambrose, cited, 140, 145.
St. Augustine, cited, 145.
St. George, bank of, 99, 100.
St. Isidore, 140.
St. John, cited, 145.
St. Peter, 153, 154.
San Domingo, see *Española*.
Sandy Point, 117.
San Lucar, 112.
San Salvador, island of, 35, 53.
Santa Maria, caravel, 30, 45.
Santa Maria de la Concepcion, island of, 35, 53, 103, 104.
Santangel, Luis de, Letter to, 52.
Santiago, 85.
Scotus, cited, 140.
Seneca, cited, 144.
Seras, 135.
Serpent's Mouth, 138.
Sierra Leone, 115, 131, 136.
Simon El Braso, cited, 145.
Solomon, 110, 230.
Sopora, Mount, 110.
Spain (*see also Castile*), 16, 27, 37, 44, 55, 60, 65, 83, 107, 110, 125, 130, 144, 146, 148, 153, 158, 163, 171, 174, 201, 204, 222; People of, 173, 215; Sovereigns of, (*see also Ferdinand and Isabella*), 27, 33, 67, 75, 83, 105, 148, 153, 158, 199, 204; Colonists from, 155, 165; Claim to America, 172.

Strebo, cited, 12, 140.
St. Vincent, cape, 112, 134, 208.
Tangier, 111.
Taprobana, island of, 110.
Theopompus, cited, 12.
Tigris, 139.
Torres, Antonio de, 172.
—Juana de la, Letter to, 151.
Tortosa, 207.
Trade, Eastern, 14; with natives, 40, 57.
Trinidad, island of, 116, 119, 121, 129, 136.
Valencia, 116, 136.
Venice, 207, 230.
Veragua, 205, 207, 211, 212, 228, 230, 232.
Verd, cape de, islands, 83, 90, 113, 131, 136.
Virgil, cited, 12.
Western lands (*see also Indies*), Suggestions of, 12; Rumors of, 19, 50, 65; Mapping of, 21; Value, 19, 22; Conditions needed for discovery of, 13; Pre-Columbian findings of, 18, 20.
West Indies (*see also Indies*), Discovery, 33, 52, 84, 107; Naming, 35, 52; Description, 37, 54; Climate, 37, 47, 55, 113, 132; Natives (*see Aborigines*); Settlements in (*see Española*); Gold in, 38, 56, 61, 68, 108, 111; Size of, 60; Colonists of, 45, 155, 165; Government of, 67, 159, 180.
World, shape of, 12, 133, 137; Size of, 16, 208.
Xaragua, 157, 160.
Yanez, Vincent, 156.
Zacharias, cited, 145.

www.ingramcontent.com/pod-product-compliance
Lightning Source LLC
Chambersburg PA
CBHW031730230426
43669CB00007B/307